# Synthesis Lectures on Mobile & Pervasive Computing

**Series Editor**

Mahadev Satyanarayanan, Carnegie Mellon University, Pittsburg, USA

This series publishes short books on mobile computing and pervasive computing, which represent major evolutionary steps in distributed systems. Although many basic principles of distributed system design continue to apply, four key constraints of mobility have forced the development of specialized techniques. These include: unpredictable variation in network quality, lowered trust and robustness of mobile elements, limitations on local resources imposed by weight and size constraints, and concern for battery power consumption. Beyond mobile computing lies pervasive (or ubiquitous) computing, whose essence is the creation of environments saturated with computing and communication, yet gracefully integrated with human users. A rich collection of topics lies at the intersections of mobile and pervasive computing with many other areas of computer science.

Peter Steenkiste

# Introduction to Wireless Networking and Its Impact on Applications

Peter Steenkiste
Departments of Computer Science
and of Electrical and Computer Engineering
Carnegie Mellon University
Pittsburgh, PA, USA

ISSN 1933-9011 ISSN 1933-902X (electronic)
Synthesis Lectures on Mobile & Pervasive Computing
ISBN 978-3-031-27465-7 ISBN 978-3-031-27466-4 (eBook)
https://doi.org/10.1007/978-3-031-27466-4



This Springer imprint is published by the registered company Springer Nature Switzerland AG
The registered company address is: Gewerbestrasse 11, 6330 Cham, Switzerland

*To my wife, children, and parents*

# Contents

# About the Author

**Peter Steenkiste** is a Professor of Computer Science and of Electrical and Computer Engineering at Carnegie Mellon University. He received the degree of Electrical Engineer from the University of Gent in Belgium in 1982, and the MS and Ph.D. degrees in Electrical Engineering from Stanford University in 1983 and 1987, respectively. His research is in networking, including many aspects of the Internet and wireless networking.

# 1 Introduction

## 1.1 Goals and Intended Audience

The Internet has had a tremendous impact on many aspects of our lives, and one reason is the availability of high performance, low cost wireless internet access. For example, WiFi provides pervasive Internet access for a variety of applications throughout our home, office, and other environments. Cellular similarly provides virtually ubiquitous Internet access, even while we are mobile. Without wireless, Internet access would much more limited. Unfortunately, achieving good performance over wireless links is much more challenging than over wired links. The reason is that wired signals travel though a carefully engineered transmission medium, while wireless signal travel through our physical environment, which is much more challenging, especially as wireless devices, people and objects move around.

Let us use a mobile user, Bob, who walks from a university campus to a coffeeshop (Fig. 1.1) as an example. Along the way, Bob uses his cellphone for a variety of tasks, e.g., reading e-mail or participating in a video conferencing call. He has the option of using the campus and coffee shop WiFi networks, or cellular. How do these options compare in terms of network performance? What is the impact on the applications? Finally, how does the performance compare, for example, with using Ethernet or WiFi in the office?

Another example might a researcher, Mary, who is working on supporting autonomous driving by offloading compute-intensive tasks to the cloud. What network topologies should she use, WiFi or cellular, and for WiFi, what version of WiFi, e.g., 802.11ax or 802.11ay? Many researchers and product developers in mobile and wireless computing need to make similar decisions.

The above questions are not easy because the answers depend on many factors, besides the network technology used. Examples include the physical environment in which the wireless devices are used (indoors, outdoors in urban or rural environments, etc.), mobility of not just the wireless devices but also other objects and people in the environment, other wireless



P. Steenkiste, *Introduction to Wireless Networking and Its Impact on Applications*,
Synthesis Lectures on Mobile & Pervasive Computing,
https://doi.org/10.1007/978-3-031-27466-4_1

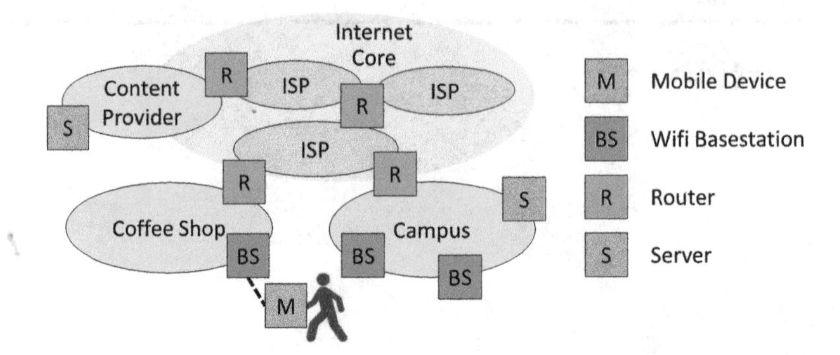


**Fig. 1.1** Mobile user accessing server on remote campus (Stick figure reused from https://svgsilh.com/svg/304880-9c27b0.svg under Creative Commons license CC0)

activity in the area, etc. The primary goal of the lecture is to teach how wireless networks, specifically wireless access links, impact the performance of applications that rely on the network. More specifically, this lecture teaches *how* wireless networks work, *why* various factors impact wireless network performance at the application level, and *what* both network engineers and application developers can do to best cope with these challenges.

In addition to users and developers of applications for wireless and mobile devices, the lecture can also be used by students and researchers who want to fill in gaps in their background. For example, a student may have completed an introductory course on networking and the Internet. Such a course often does not cover wireless, but the student can use the lecture to learn how wireless access links impact the protocols covered in the networking course. Similarly, a student studying wireless communication may be interesting in learning how decisions made in the wireless physical layer impact the rest of the system.

The lecture assumes that the reader has a good background in computer science and in how computer systems work. Each chapter provides background, at a high level, that is relevant to the contents of the chapter. We also provide references to textbooks and papers that provide more detail.

## 1.2 Structure of the Internet

The Internet is sometimes referred to as a "network of networks". It consists of large number of Internet Service Providers (ISPs) (shown as "Internet Core" in Fig. 1.1) that provide Internet connectivity to a set of "stub networks" at the edge of the Internet (the Campus and Coffee Shop networks in our example). Other examples of stub networks are Internet content and service providers. Communication across the Internet is made possible by a set of *protocols* that support communication inside each network (ISP or stub network), across the Internet core, and end-to-end between the two communicating devices.

In this section, we first introduce the Internet *protocol stack*, which includes all the protocols a device needs for end-to-end communication over the Internet and we explain how it supports end-to-end communication. We also discuss the benefits of the modular protocol stack design.

### 1.2.1 The Protocol Stack

Internet communication is implemented as a set of *protocols*. Broadly speaking, a protocol is a set of rules that allows two parties to communicate. In the Internet, the rules cover the format of packets, interpretation of the packet contents, and a set of actions that devices must take when they receive a packet. Each device in the Internet, including endpoints and devices inside the network, implement the protocols they need as a set of modules that are organized as a 7-layer stack, based on the Open Systems Interconnect (OSI) model.

Figure 1.2 shows the protocol stacks for the various devices in the scenario in Fig. 1.1. To simplify the figure, we only include the five layers that are relevant to the lecture. We also assume that the networks for the two endpoints connect to the same ISP. We see that the two endpoints implement all the layers in the stack. Routers are responsible for forwarding packets between networks (ISPs or stub networks) using a network protocol and they only implement the bottom three layers. Finally, basestations, and more generally switches, forward packets between links inside a network using the datalink layer. They only implement the bottom two layers. Note that in practice, routers are often also used inside individual networks for scalability and network management purposes.

The horizontal arrows in Fig. 1.2 represent *protocols* that allow two modules in the same layer on different devices to communicate. Protocols can be end-to-end (between devices and applications), inter-domain (used by routers to connect networks), and intra-domain (inside individual networks). Protocols make use of *protocol headers* that are inserted in

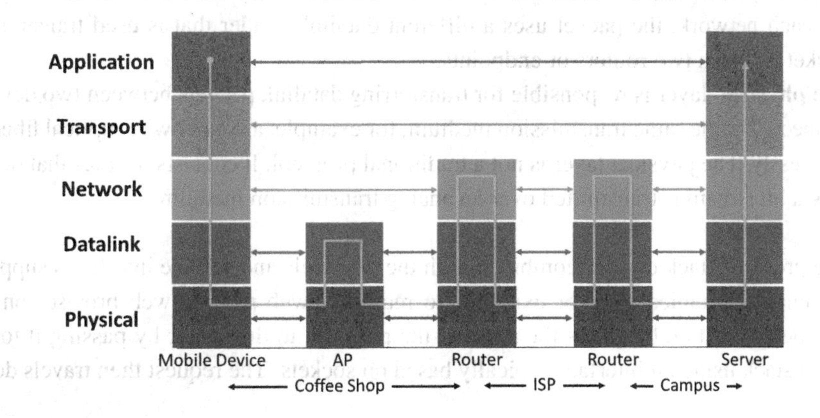


**Fig. 1.2** The network protocol stack

the packet by the sending module to provide information the receiving module needs to interpret and act on the data. For example, when the mobile device wants to download a web page from a web server on the campus, its request includes an HTTP GET header, while the server's response would include an HTTP Response header. Both headers include additional information, such as the requested URL or the web page.

The vertical interfaces between adjacent modules on the same device are *service interfaces*. For example, the transport layer provides a service to the application layer. To provide this service, the transport layer requires the use of the network layer service, etc.

Let us now briefly review the responsibilities of each layer. More details will be presented as needed in later chapters:

- **Application** protocols are used by distributed applications, often based on a client-server paradigm. Many application protocols have been standardized, for example HTTP (used for web browsing) and SMTP (used for e-mail). Of course, users can also implement their own custom protocols.
- **Transport** protocols are end-to-end protocols that ensure that packets are delivered to the right application on the destination host. They can also provide additional services such as reliable, in order delivery of data as supported by TCP.
- **Network** protocols are responsible for delivering packets to devices across the Internet. Today, the Internet uses the Internet Protocol (IP). Most traffic uses IP version 4, but an increasing fraction of the traffic uses version 6. Packet delivery is based on an IP address, which is 32 bits for IP v4 and 128 bits for IP v6. The IP protocol supports a minimal service: best effort packet delivery. This means that packets may be lost, reordered, or even duplicated. Most applications rely on a transport protocol such as TCP for reliable data delivery.
- **Datalink** protocols implement a specific network technology such as Ethernet, WiFi, or LTE. As a packet travels end-to-end, for example between the mobile device and server in our example, it often traverses multiple networks, which may use different technologies. On each network, the packet uses a different datalink header that is used transmit the packet between two routers or endpoints.
- The **physical** layer is responsible for transferring datalink packets between two devices connected to the same transmission medium, for example, a copper wire, optical fiber, or wirelessly. The physical layer is not a traditional protocol. It consists of rules that define how a bit stream is transmitted over an analog transmission medium.

The protocol stack design, combined with the protocols and service interfaces supports end-to-end communication. For example, to request a web page, a web browser on the mobile device in Fig. 1.2 sends the request, the payload, to the server by passing it to the protocol stack using an interface typically based on sockets. The request then travels down the protocol stack (yellow arrow in the figures), with each protocol prepending a header to the payload. At the bottom of the protocol stack, the physical layer transmits the payload, including the headers, to the next device in the path to the destination.

This device, a basestation in our example, passes the payload up the stack. In each layer, the protocol extracts "its" header, i.e., the one inserted by it peer, acts on it, and then passes the payload to the next layer in the stack. This process repeats itself until a protocol finds that it is the destination identified in the header. In our example, the mobile device will have used the address of the basestation in the datalink header, so the datalink module in the basestation knows it is responsible for forwarding the packet. It then determines what device should receive the packet next, attaches the appropriate header, and hands the payload down the stack. This process repeats itself until the packet reaches the destination device, where the payload is passed up to the intended application, in our example the web server.

### 1.2.2 Benefits of the OSI Model

The OSI model restricts how developers can implement networking code. While this may seem rigid, it has some significant advantages.

First, an architecture based on per-device protocols stacks is modular design with well defined horizontal (protocols) and vertical (service) interfaces, so the implementation of the individual modules can be different across the Internet and can be changed easily. This is important because many companies and developers build network hardware and software components and these many parts need to interoperate and work together on a very large scale.

Second, how protocols at each layer operate is not relevant to the other layers, so it is easy to introduce new protocols without impacting the protocols in other layers. As long as the new protocol respects the existing service interfaces, a host using the new protocol will be able to communicate with other hosts that also support the new protocol. This is possible because the protocol headers and service interfaces include demultiplexing information that makes sure that packets are handed to the correct protocol in each layer as payloads are passed up or down the stack.

In practice, the diversity in the layers is very high at the top (applications) and near the bottom (physical and datalink protocols), but it is low in the center of the stack. For example, at the network layer only IP v4 and v6 are used and they are very similar. The reason is that a network protocol requires very widespread support throughout the Internet to be useful, while others protocols can be used with more limited levels of deployment.

## 1.3 Outline of the Lecture

The core Internet technology was developed as part of the ARPAnet (Robert, 1986; Cerf & Kah, 1974) project and both the ARPANET and the early Internet were entirely wired. The first wireless packet data network was the ALOHAnet (Schwartz & Abramso, 2009; Abramson, 1970), which is widely viewed as a predecessor of both Ethernet and WiFi. However, it was not until the introduction of WiFi in the1990s, more than 20 years later, that the use of wireless access links in production networks became feasible.

Given the modular nature of the protocol stack, replacing wired links by wireless links should have been transparent to the rest of the protocol stack. Unfortunately, that is not the case. Wired links are engineered for high performance. Their throughput is fixed and known, since the transmission medium (copper wires or optical fiber) is standardized. In contrast, wireless signals travel in free space, which is a more challenging transmission medium, and performance is highly variable, depending on the physical environment. It is not always possible to hide these differences from higher-level protocols. For example, links often have higher latency, and lower and more variable throughput. Wireless connection may also not always be available. Finally, wireless access links allow mobile devices to be connected to the network, raising another set of challenges.

In this lecture we introduce the unique challenges associated with wireless Internet access and we describe techniques that can reduce their impact on applications. Similar to many networking textbooks, this lecture describes the protocol in the layers in a bottom-up fashion. However, instead of presenting a general introduction to protocols, each chapter first provides background on widely used protocols and technologies in that layer, and it they describes (1) how the protocols and technologies are impacted by the use of wireless links, and (2) techniques that can be used to address wireless challenges.

The bottom two layers of the stack are specific to the network technology used, so in wireless networks, they can be engineered specifically for wireless. In Chap. 2 we describe at a high level how wireless signals propagated and the challenges they face in different physical environments. In Chap. 3 we discuss physical layer techniques that can improve the throughput of wireless links. In Chap. 4 we present some background on datalink protocols and then describe the protocols used in WiFi and cellular, two technologies that focus on maximizing throughput in the unlicensed and licensed spectrum respectively.

Higher layer protocols are used throughout the Internet (network layer) or end-to-end (transport layer) so they cannot easily be optimized for wireless links. Fortunately, today's WiFi and cellular physical and datalink layers are very sophisticated and generally have good performance. Nevertheless, performance properties such as lack of wireless connectivity and differences in latency and throughput cannot be hidden. In Chap. 5 we look at several wireless and mobility challenges related to the Internet protocol and transport protocols, specifically reliable streaming protocols such as TCP.

Finally in Chap. 6, we discus how applications on wireless and mobile devices are affected by the presence of wireless links and present techniques that can reduce the impact. We focus on three classes of challenges. First, while wired Internet access is very reliable, wireless and mobile users may be disconnected at certain times. Second, wireless links often have lower and more variable throughput compared with wired access links. Finally, mobile devices are resource constrained so supporting compute intensive applications is challenging.

# The Physical Layer: Sending Bits

# 2

## 2.1 The Wireless Spectrum

We discuss key properties of the main spectrum bands that are widely used for personal communication.

### 2.1.1 Wireless Spectrum Use Examples and Trends

The RF spectrum ranges from a few KHz to 300 GHz. It supports a wide variety of wireless communication technologies. In addition, some wireless communication technologies use the infrared and visible light bands (300 GHz–380 THz range). The technologies we discussed in this lecture mostly use frequencies in the 900 MHz–60 GHz range. While this may seem like a lot of spectrum, spectrum properties differ a lot across this range and not all bands are equally attractive. In addition, there is a lot of demand for spectrum, not just for personal wireless communication (e.g., cellular and WiFi) but also for other types of communication (satellite, mobile first responders, ...). Spectrum is also used for purposes other than communication, e.g., radar, GPS, astronomy research, etc. In short, spectrum is a precious resource.

Table 2.1 identifies the frequency ranges of the RF spectrum that are widely used for wireless data communication. AM radio uses the Medium Frequency (MF) band, while FM radio uses the Very High Frequency (VHF) band. Broadcast TV channels can be found in the VHF and Ultra High Frequency (UHF) bands. These lower frequency bands also support many mobile technologies, including Land Mobile Radios (LMR) used by first responders and some of the bands used by cellular operators.

The UHF, SHF and EHF spectrum ranges include several unlicensed bands that are widely used by consumer devices. First, the 900 MHz band is used by many IoT protocols, while



P. Steenkiste, *Introduction to Wireless Networking and Its Impact on Applications*, Synthesis Lectures on Mobile & Pervasive Computing,
https://doi.org/10.1007/978-3-031-27466-4_2

**Table 2.1** Spectrum ranges

| Designation | Frequency range | Wavelengths |
|---|---|---|
| Low frequency | 30–300 KHz | 10 km–1 km |
| Medium frequency | 300 KHz–3 MHz | 1 km–100 m |
| High frequency | 3–30 MHz | 100 m–10 m |
| Very high frequency | 30–300 MHz | 10 m–1 m |
| Ultra high frequency | 300 MHz–3 GHz | 1 m–100 mm |
| Super high frequency | 3–30 GHz | 100 mm–10 mm |
| Extremely high frequency | 30–300 GHz | 10 mm–1 mm |

the 2.4 GHz band is used by WiFi, Bluetooth and other wireless technologies. The Super High Frequency (SHF) band includes the 5 GHz band and the recently allocated 6 GHz band, which are also used by WiFi and other technologies. Finally, the Extremely High Frequency (EHF) spectrum, often referred to as the "mmWave" band, is used by some of the most recent versions of WiF and 5G cellular. The above overview ignores many wireless communication technologies that will not be discussed in this lecture. Examples include satellite communication, communication in the optical frequency band, and many older technologies used by government and other organizations.

When we consider the use of the spectrum for communication over time, we can observe several trends. Early technologies operated in the lower frequencies (below 1 GHz) and often used analog technologies. This is consistent with the state of the technology at the time. Operating at higher frequencies requires faster circuits, which are more challenging to build. The technologies also predate the Internet and wireless communication used analog technology, e.g., broadcast radio and TV, etc. Over time, technologies improved, making it possible to use higher frequency bands in commodity devices. This was necessary to meet the growing demand for spectrum. It also enabled a switch to digital technologies, which offers many benefits. It allows error checking and correction (quality), compression (performance), and encryption (privacy). Finally, over time, wireless technologies continued to improve, resulting in higher throughputs and a more efficient use of the spectrum (in bits/sec per Hz).

### 2.1.2 Spectrum Allocation

The allocation of spectrum is decided by individual countries, often in collaboration with other countries through organizations such as the International Telecommunications Union (ITU). In the US, allocation is done by the Federal Communications Commission (FCC) and the National Telecommunications and Information Agency (NTIA). The NTIA is responsible for spectrum allocation for federal government communication while the FCC is responsible for spectrum allocation to commercial entities and state and local governments. A detailed chart showing the allocation of frequency bands for the US can be found online (National Telecommunications and Information Administration, 2016).

Broadly speaking, frequency bands fall in two categories, bands for licensed and unlicensed use. We will refer to them as licensed and unlicensed bands or frequencies. This is somewhat of an approximation since some bands can have both licensed and unlicensed users, as discussed below.

For licensed spectrum, using the band, e.g., to transmit RF signals, requires a licensed assigned by the FCC or NTIA. Well known examples are spectrum bands allocated to radio and television stations for over-the-air broadcast and to cellular operators to support phone calls, mobile Internet access, and other services. In contrast, unlicensed spectrum can be used by any user, assuming they use a device that meets the FCC requirements for the spectrum band. These requirements place constraints on the RF transmissions that are allowed, such as limits on the transmit power, constraints on how bits are transmitted, and what type of datalink protocols can be used. These constraints are designed to allow multiple independent users to share the spectrum in a reasonably efficient way. Practically this means that wireless radios, for example for WiFi, must be tested and certified before they can be sold for use in the unlicensed spectrum.

Some unlicensed spectrum bands can also be used by applications other than communication. For example, microwave ovens use the 2.4 GHz unlicensed band, which is also used by many wireless communication technologies (Bluetooth, WiFi, cordless phones, baby monitors, etc.).

Finally, some bands can have multiple licensed users, or can have both licensed and unlicensed use. One example is cases where users only need access to spectrum in certain geographic areas. For example, spectrum for maritime radar or for airport communication is only used by the license holder near airports and in coastal regions. In that case, other users, licensed or unlicensed, can be allowed to use the spectrum in other areas, under certain conditions. Also, many licensed bands have very low utilization so allowing unlicensed users (referred to as secondary users) to use the band without interfering with the license holder (the primary user) increases the spectrum band's utility. We present some examples in the next section.

### 2.1.3 Licensed Versus Unlicensed Spectrum: Implications for Protocols

Whether a band is licensed or unlicensed, or has only primary or also secondary users, has an impact on the type of protocols that are needed.

A licensed band in a geographic region is typically allocated to a single license holder, e.g., a TV station or a cellular operator. This means that the license holder has full control over all transmissions and it does not have to worry about competing transmissions from third parties. This has two significant advantages. First, it simplifies the datalink layer protocol since it does not have to coordinate with third parties that operate independently. Second, it is easier for the license holder to provide commitments to users about the "Quality of Service" (QoS) it provides. As a result, it has complete control over how it allocates bandwidth across its users, e.g., based on priority, payment plan, etc. In Sect. 4.4 we discuss cellular protocols as an example of a technology that uses licensed spectrum.

In contrast, unlicensed spectrum band are shared by many independent users. As a result, protocols will be designed in such a way that independent users can efficiently share the spectrum, both when using the same protocol (e.g., 802.11n), or different protocols or protocol versions (e.g., 802.11n, 802.11ax, and Bluetooth in the 2.4 GHz band). This requires protocol features that allow them to co-exist, which introduces overhead. This can be challenging and does not always work well when the protocols that very different, e.g., Bluetooth or low-power protocols in general and WiFi. An even more extreme example is microwaves. Understandably, microwaves simply radiate energy and they do not follow any protocol, so any energy that leaks from the enclosure will interfere with nearby communication in the 2.4 GHz band. In Sect. 4.3 we discuss WiFi protocols as an example of a technology that uses unlicensed spectrum.

### 2.1.4 Supporting Secondary Users

Frequency bands with a primary license holder and one or more secondary users require protocol support that ensures that the primary user is not affected by the presence of the secondary users. Since primary users generally use protocols designed for use in licensed spectrum bands, they do not coordinate with other users, so it is the responsibility of the secondary users to avoid interfering with the primary users. Let us consider two examples.

TV stations uses the spectrum in the 54–806 MHz range. This a lot of sub-GHz spectrum that is very attractive because it has relatively low attenuation. However today, relative few people watch over the air broadcast TV and the number of over-the-air TV channels has been decreasing in many cities. In 2008, the FCC approved the unlicensed use of *TV white spaces*, TV channels that are not being used for TV transmission. In order to avoid that secondary users interfere with TV transmissions, the FCC mandated the use of a *spectrum database*. TV stations are required to register their TV station and its location and transmission range in the spectrum database and unlicensed users that want to use the TV white spaces are required to check the database for before using a TV channel.

Another example is the Citizens Broadcast Radio Service (CBRS) which uses the spectrum band of 3.5–3.7 GHz. The FCC defines three tiers of users for this band. First, *incumbent users*, which includes some satellite ground stations and the US Navy, were assigned an exclusive license a long time ago and continue to have exclusive access in some areas. Second, in 2020 some of the unused spectrum in some areas was auction off to *Priority Access Licensees* (PAL), mostly cellular operators. They have an exclusive license for the spectrum they acquired. Finally, *General Authorized Access* (GAA) users can use the spectrum for free, as long as they do not interfere with incumbents or priority access licensees. Potential

GAA users must request, and be granted, access to CBRS spectrum before they can use it. GAA users can for example deploy a private cellular network for research purposes (CMU, 2020). To avoid interference, the FCC established the Spectrum Access System (SAS) which controls access to unused CBRS spectrum. It operates an Environmental Sensing Capability (ESC), sensors that track the use of CBRS spectrum.

### 2.1.5 Mobile Users

Decisions on spectrum use are made by individual countries, but mobile devices may use their radios in multiple countries, e.g., the mobile devices may be sold worldwide, or users may use their devices when they travel internationally. This raises the question is how a device can find out what frequency bands it can use in a particular location. Different solutions are used for different spectrum bands.

The International Telecommunications Union (ITU) has defined several Industrial, Scientific and Medical (ISM) bands that can be used by unlicensed devices world-wide. Examples include unlicensed bands at roughly 900 Mz, 2.4 GHz, 5 GHz and more recently, 60 GHz. The ISM bands are very attractive for wireless technologies such as WiFi and Bluetooth, since mobile devices can access networks anywhere. While ISM bands are defined globally, many details of the ISM bands and their use are specified in national radio regulations, so they can differ across countries. Examples are the precise frequency ranges of the bands and the policies for using the spectrum can differ. Protocols standards must provide mechanisms that allow mobile devices to comply with the local spectrum polices. For example, for WiFi, the local infrastructure (e.g., WiFi basestations) can inform devices about local policies.

Cellular technologies face a similar challenge. The frequency bands used by a cellular operator differs across regions, even in the same country. In addition, large cellular operators often have agreements with operators in other countries that allow their customers to use their phones in those countries, for example using an international plan. To allow cellular devices to discover what local cellular networks they can use in a specific area and what frequency bands they use, operators download information on the devices of their customers about both the operator's network and networks they have agreements with. When switched on, the cell phone first try to connect to its "home network" using the pre-stored information. If that fails, the devices tries to connect to the other networks listed.

## 2.2 Wireless Communication Basics

Wireless communication is based on the transmission of an *electro-magnetic signal* from a transmitting device to a receiving device (Fig. 2.1). An electro-magnetic (EM) signal consists of an electric-magnetic field that is transmitted by a sender through an antenna and picked up

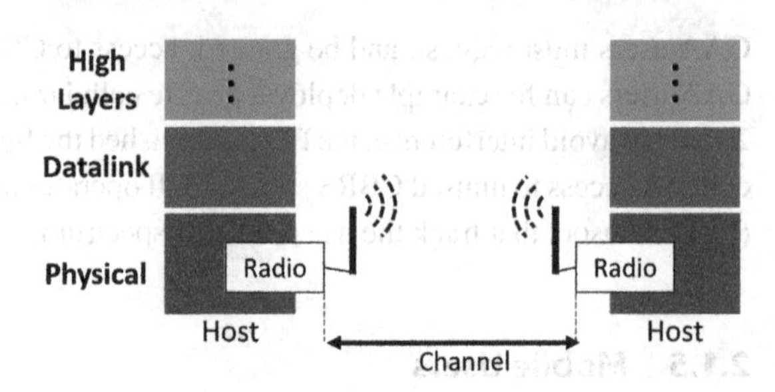


**Fig. 2.1** Communication over a wireless channel

by the receiver's antenna. EM signals travel at the speed of light ($3 * 10^8$ m/s) in a vacuum. The signal propagation speed is lower in copper wires and optical fibers.

Communication is achieved by a process called *modulation*: the sender changes one or more properties of a reference EM signal in a way that reflects the values of the bits in the packet it is sending to the receiver. The transmitter and receiver synchronize their clocks on the reference signal at the start of the packet transmission. The receiver *demodulates* the signal to retrieve the bit values. Specifically, it analyzes the received signal to identify the changes the sender made to the reference signal and uses the observed changes to reconstruct the values in the transmitted bit stream. Modulation and demodulation are implemented by a wireless radio. Most wireless radios can both transmit and receive although there are exceptions, e.g., over the air TV broadcast. In this lecture we focus on digital communication but modulation of EM signals can also be used to transmit analog information. Examples are over the air AM and FM radio broadcast.

The reference signal that is used for wireless communication is called a *carrier signal* and it can be thought of as a sine wave:

$$A_c \times sin(\phi_c + f_c \times t) \tag{2.1}$$

where $A_c$ is the amplitude, $f_c$ is the frequency and $\phi_c$ is the phase of the carrier signal at time $t$. They represent the three signal properties that can be used for modulation. Specifically, the transmit radio modules the carrier signal by changing its amplitude, frequency, and/or phase based on the data that is transmitted. Changing the phase corresponds to a small shift in time of the signal.

When transmitting a bit stream to another device the transmitter modulates a sequence of *symbols*, each representing one or more bits. Figure 2.2 shows some examples of *binary* modulation, where the transmitter sends one bit at a time using a carrier signal with amplitude $A_c$, frequency $f_c$, and phase $\phi_c$. The three examples show how the sender can change the amplitude (between $A_c$ and 0), the frequency (between two frequencies $f_1$ and $f_2$ with a different offset relative to $f_c$), or the phase (0 and 180 degree offsets relative to $\phi_c$). Note that the figure is not to scale. For example, in this lecture we will discuss technologies that use carrier signals between 90 MHz and 60 GHz, while symbol rates are several orders of

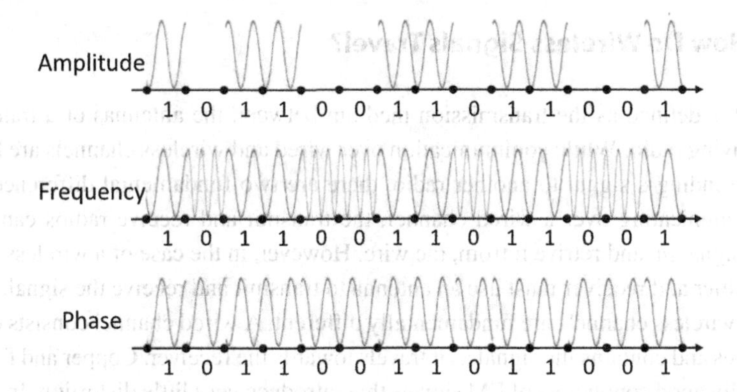


**Fig. 2.2** Examples of binary modulation (Sine wave from https://www.wisc-online.com/assetrepository/viewasset?id=4107 by "unknown". Reused under license CC-BY-NS 2 (https://creativecommons.org/licenses/by-nd/2.0/))

magnitude lower. Today's wireless technologies use multi-bit symbols since they make it possible to transmit at higher bit rates (see Sect. 3.1). For example, 802.11ax can use symbol sizes up to 10 bits.

Effective communication requires a protocol standard that specifies, at a minimum, the frequency of the carrier signal and how the information that is communicated (symbols values in our case) is mapped onto changes to the carrier signal.

Figure 2.3 shows how a transmitter transmits a modulated EM signal using an antenna. The signal will travel in many directions (see Sect. 3.7) so potentially several receivers will be able to pick up a copy of the signal using their antenna. The EM signal arrives at the receiving antennas with a delay $c \times d$, where $c$ is the speed of light and $d$ is the distance between the transmit and receive antennas. As it travels through the space between the two antennas, the EM signal is attenuated and distorted so the received copy is not identical to the transmitted signal, so it can be very challenging for the receiver to correctly demodulate the received signal. The goal of this chapter is to explain these challenges. We present several techniques to address them in the next chapter.

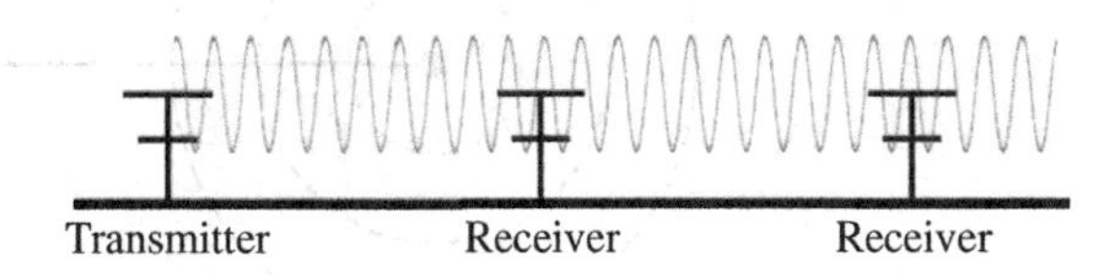


**Fig. 2.3** End-to-end view of a wireless signal (Sine wave from https://www.wisc-online.com/assetrepository/viewasset?id=4107 by "unknown". Reused under license CC-BY-NS 2 (https://creativecommons.org/licenses/by-nd/2.0/))

## 2.3 How Do Wireless Signals Travel?

A *channel* is defined as the transmission medium between the antennas of a transmitting and a receiving radio. While communication over wired and wireless channels are based on one radio sending a signal to another radio, there are two fundamental differences. First, when communicating over a wired channel, the transmit and receive radios can directly inject the signal in, and retrive it from, the wire. However, in the case of a wireless channel, the transmitter and receiver must use an antenna to transmit and receive the signal. Second, wired and wireless channels are fundamentally different. A wired channel consists of a wire that captures and contains the signal as it travels towards the receiver. Copper and fiber links are generally good conductors of EM signals that introduce very little distortion. In contrast, with wireless communication, the signal travels through the physical environment, which introduces many challenges as described in detail in Sects. 2.5–2.7.

In this section we introduce three simple abstractions that are sufficient to explain both the wireless challenges associated with wireless communication and the techniques that can be used to reduce their impact on performance.

### 2.3.1 The "Energy" View

The first view is the "energy view". An EM signal represents energy that is radiated into free space by the transmit antenna. In the simplest case, the energy radiates in each direction with the same signal strength (energy level expresses in Watt), as shown in Fig. 2.4. Since the EM signal travels at the speed of light, the transmitted signal can be visualized as a sphere with the signal's energy evenly distributed over the surface of the sphere. The sphere is centered at the transmit antenna and the radius of the sphere increases at a rate of $3 \times 10^8$ m/s which means that the amount of energy that a receiving antenna receives decreases with the distance $d$ (see Sect. 2.5). An antenna that transmits energy evenly in all directions

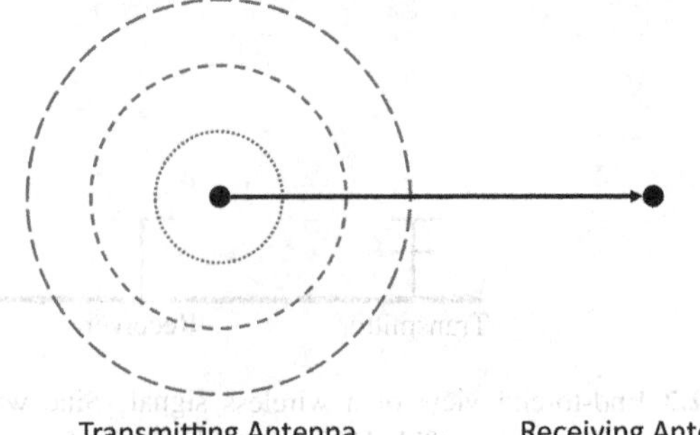


**Fig. 2.4** Wireless signals as energy

is called an *isotropic antenna*. In practice, antennas are designed to focus the transmitted energy in specific directions (see Sect. 3.7).

### 2.3.2 The "Ray" View

The simplest signal propagation environment is a *free space* without objects as illustrated in Fig. 2.4. In this case, the receiving antennas will only be able to capture the energy that is sent directly towards it. We can view the energy that travels between the two antennas as a "ray" that follows a Line of Sight (LoS) path between the two antennas.

The "ray view" is more interesting in environments with obstacles, such as the ground, walls, furniture, people, etc. In that case, energy sent in directions other than the LoS path can also reach the receiver if it is reflected by these objects. For example, in Fig. 2.5a there is no LoS path between the two antennas, but the two devices may still be able to communicate using the signal that is reflects off a wall. Because of reflections, receivers can also receive multiple copies of the signal (Fig. 2.5b). This is analogous to having echoes when sound is reflected by multiple surfaces in a large, enclosed space, so a listener hears multiple attenuated copies slightly offset in time. We discuss the impact of multi-path on wireless communication in more detail in Sect. 2.6.1.

### 2.3.3 The "Frequency" View

The energy and ray models present a time domain view of wireless signal propagation (Fig. 2.3). However, there are several reasons why we also need a "frequency view" of wireless signals. A first reason is that the FCC mandates that specific frequency bands can only be used by specific users for specific types of communication. This means that for each frequency band, it is the responsibility of its users to make sure that the wireless signals they transmit are confined to the frequency band allocated to them, which is difficult to do in the time domain. Another example is that some wireless communication challenges are

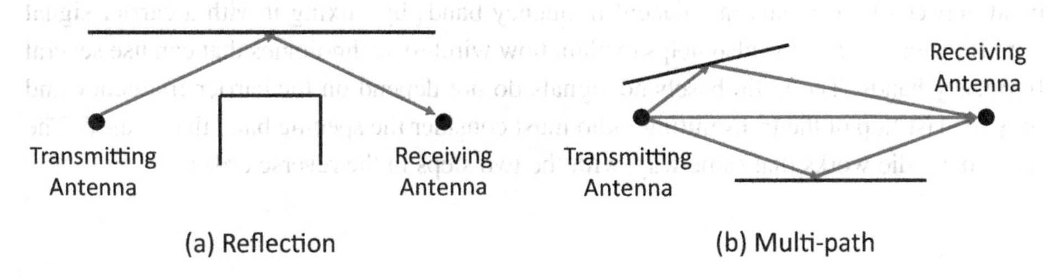


**Fig. 2.5** Wireless signals as a ray

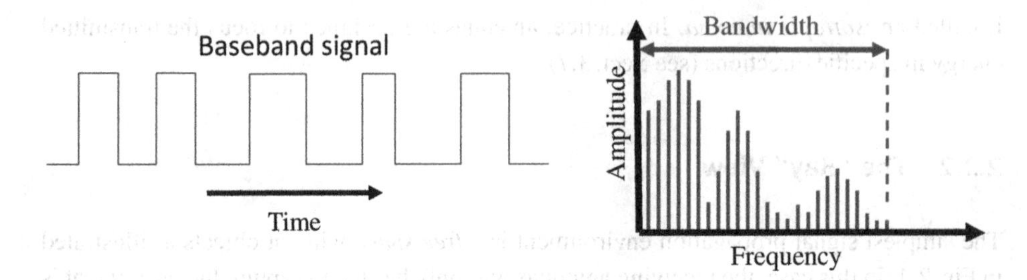


**Fig. 2.6** Frequency view of wireless signals

*frequency selective* in the sense that their impact depends on what part of the spectrum is used. These requirements and properties cannot be enforced or modeled using a time domain view.

The frequency view of the signal is the result of the fact that any periodic signal can be represented as a sum of sine wave with different frequencies, amplitudes, and phases.[1] One way of visualizing the signal in the frequency domain is to show how the signal's energy is distributed across the frequency spectrum, as illustrated in Fig. 2.6. The frequency view shows us, for example, how much spectrum the signal uses and in what part of the spectrum, e.g., is a 20 MHz WiFi signal properly aligned in the 2.4 GHz unlicensed band. Similarly, we can model frequency-selective channel effects by applying attenuation and phase shifts to each of the (digitized) frequencies in the signal. Translating between a signal's representation in the time and frequency domain can be done digitally using a Fast Fourier Transform (FFT), a well-known algorithm.

Finally, the frequency view helps in understanding how many radios can share the spectrum without interfering with each other, even when they are in the same geographic area, radios generate the modulated "baseband" signal they want to transmit as a low-frequency signal, such as the signal in Fig. 2.6. They then "mix" this "baseband" signal with a carrier signal with a frequency that corresponds to the center frequency of band that the radio is supposed to use. Mixing effectively "shifts" the modulated signal to the target frequency band. For example, Fig. 2.7b shows how the signal from Fig. 2.6 is placed in a frequency band between two signals in adjacent frequency bands by mixing it with a carrier signal with frequency $CF$. This also helps explain how wireless technologies that can use several frequency bands. The radio baseband signals do not depend on the carrier frequency and only the last step of the transmitting radio must consider the specific band that is used. The receiving radio works that same way, with the two steps in the reverse order.

[1] Signals used for wireless communication are not periodic, but this does not impact the frequency view model.

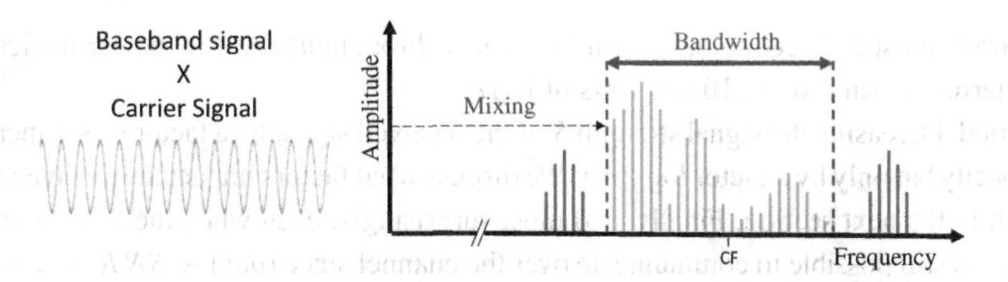


**Fig. 2.7** Wireless communication in a specific frequency band (Sine wave from https://www.wisc-online.com/assetrepository/viewasset?id=4107 by "unknown". Reused under license CC-BY-NS 2 (https://creativecommons.org/licenses/by-nd/2.0/))

## 2.4 Fundamental Limit of Throughput

The primary network metric of interest for a wireless link is its capacity, i.e., how many Mbits/second of bandwidth can it support. Link capacity depends on my factors including the quality of the radios and the physical environment through which the wireless signal travels. Over the years, the wireless communication community has been able to dramatically increase the amount of bandwidth that can be supported by a given wireless channel. This raises an interesting question: is there a limit to the throughput that can be supported by a channel of $B$ Hz, or can improvements in radio technology be used to increase channel capacity forever. In this section present a fundamental result, known as *Shannon's theorem*, that defines an upper bound on the capacity of a channel.

The maximum capacity of a channel is defined by:

$$C = B \times log_2\left(1 + \frac{S}{N}\right) \tag{2.2}$$

where $C$ is the maximum capacity of the channel (in Mbits/sec), $B$ is the width of the channel (in Hertz), and $S$ and $N$ are the signal strength of the received signal and the noise at the receiver in the channel (in Watts). Here is an example of using Shannon's law to calculate maximum channel capacity:

**Example 2.1** Let $B$ be 20 MHz and let $\frac{S}{N}$, called the signal-to-noise ratio (SNR) be 251. Then, $C = 20 \times 10^6 \times log_2(1 + 251) = 160$ Mbps.

Shannon's law provides several insights in what factors impact channel capacity. First, the channel capacity is directly proportional to the channel width $B$. The channel width is limited by both the FCC, which allocates spectrum, and the technology used. This linear dependency is not unexpected and it explains why technologies that focus on high throughput, such as WiFi and cellular, use wide channels (e.g., 10 s of Mbps) For example, WiFi has overtime increased the channel width from 20 MHz for the early versions to 40, 80 MHz, and most recently to 160 MHz for 802.11ac and ax, as a relatively easy way to increase throughput.

In contrast, most IoT technologies that target low throughput and low power devices use much narrower bands (e.g., 10 s or 100 s of KHz).

Second, increasing the signal strength $S$ of the received signal by a factor $n > 1$ increases the capacity but only by a factor $log_2(N)$. We discuss what factors impact the received signal strength in the next section. Finally, and most surprisingly, even when the SNR is smaller than 1, it is still possible to communicate over the channel since $log(1 + SNR)$ is a positive number. Ultra Wide Band (UWB) technologies use signals that are below the noise floor. They can achieve high throughputs by using very wide channels, often several GHz wide. UWB technologies are not used much today.

Unfortunately, while Shannan's theorem defines an upper bound on the capacity of a channel, it does not informs us how to achieve this limit. It is left up to standards organizations and radio designers to figure out how to optimize channel capacity. This is a hard problem because, as described later in this chapter, the signal is attenuated and distorted in ways that depend on the physical environment as it travels to the receiver, reducing the bit rate that can be supported. In addition, the receiver needs to deal with interference from other wireless transmitters and different types of noise. Nevertheless, the radios used today by high performance wireless protocols such as WiFi and cellular come very close to the Shannon limit. We discuss the techniques they use in the next chapter.

Next, we review the main challenges at the physical layer that impact wireless network performance.

## 2.5 Signal Strength Attenuation

Since the received signal strength has a big impact on throughput (Shannon's limit), it is important to quantify the attenuation as a function of the distance between the transmitter and the receiver, and other features of the physical environment.

### 2.5.1 Free Space Communication

Equation 2.3 shows the attenuation, or "path loss", in free space, an environment without any obstacles that weaken or reflect the signal as it travels from the transmitter to the receiver:

$$Loss = (4\pi 2)^n \times \frac{f^2}{G_T G_R} \tag{2.3}$$

Let us consider the three factors impacting pass loss one by one.

Let us first consider free space loss when both endpoints use isotropic antennas. This means that the transmitted signal will propagate as an expanding sphere with the transmit antenna at the center. The signal's energy is uniformly distributed over the surface of the sphere (Fig. 2.4), which is equal to $4\pi d^2$ where $d$ represents the distance between the two devices. The attenuation caused by free space propagation is thus represented by the first denominator in Eq. 2.3 with $n = 2$.

Second, the definition of isotropic antenna used for receiving the signal specifies how much of the signal's energy the receiving antenna will pick up. In wireless communication, distances in space are often measured in terms of the number of wavelengths, rather than meters. With an isotropic antenna, the amount of energy it captures (which can be thought of as a part of the sphere's surface) increases as $\lambda^2$, where $\lambda$ is the wavelength of the carrier signal. This means that the path loss increases as $f^2$, where $f$ is the carrier signal's frequency. This effect is captured in the second enumerator in Eq. 2.3.

Third, we need to account for the fact that isotropic antennas are neither practical to build nor efficient. In practice, antennas direct most of their energy in specific directions where receivers are most likely located (see Sect. 3.7). This effect is captured by the *gains* of the actual antennas, relative to an isotropic antenna. The antenna gain in a particular direction is the ratio of the strength of the signal transmitted (or received) in that direction relative that the strength of signal transmitted (or received) by an isotropic antenna. The gain is larger than one for directions for which the antenna is optimized and less than one in other directions. Note that antennas gains are symmetric, i.e., for a given antenna and direction, the gain is the same when transmitting and receiving a signal to/from a specific direction. The values $G_T$ and $G_R$ in the denominator are the *gains* of the transmit and receive antennas.

### 2.5.2 Communication in Real Environments

In practice wireless communication does not take place in a free space environment. First, the transmission medium (the atmosphere in wireless) absorbs some of the energy. Second, and more importantly, any objects that are located between the sender and the receiver and partially or fully block the signal will absorb and reflect some of the energy, reducing the energy reaching the receiving antenna. For now we will assume that there is only single path between the transmitter and the receiver.

Calculating the exact path loss contribution of objects in a real-world environment is complicated, so its effect is typically approximated using a general formula that adjusts the result of the free space path loss equation to account for objects. The idea is that the impact of obstacles and absorption by the atmosphere have on path loss increase roughly with distance. This is captured by increasing the exponent 2 that is used in the formula for the surface of a sphere to a value $n$. $n$ is called the path loss exponent and its value depends on the properties of the physical environment such as the density and nature of obstacles. For example in an office environment with soft partitions, the value could be 2.5 while in building with solid walls it could be 3 or higher depending on the construction of the walls and floors.

Equation 2.3 offers interesting insights in how the environment impacts wireless communication and wireless link performance, but unfortunately, using it requires knowledge of the antenna gains, which may be difficult to obtain in practice. A more practical approach is to measure the path loss $Loss_0$ at a reference distance $d_0$. The path loss $Loss_d$ can then be estimated for any distance $d$ using the following formula

$$Loss_d = Loss_0 \times \left(\frac{d}{d_0}\right)^n \tag{2.4}$$

Note that the equation assumes that the antenna gains in all receiver locations are the same or similar.

### Link Budget

In summary, the impact on the received signal strength can be summarized in the link budget equation:

$$P_R = \frac{P_T \times G_T \times G_R}{L_{FS} \times L_O} \tag{2.5}$$

The power of the received signal $P_R$, equals the transmitted power $P_T$ times the gains of the two antennas ($G_T$ and $G_R$), divided by the free space loss ($L_{FS}$) and other losses ($L_O$), such as absorption by the atmosphere or objects. It is the amount of signal power available at the receiving radio for extracting bits from the received signal. Because of Shannon's limit, the link budget has a significant impact on the bit rate of the wireless link.

Similar to the audio community, the wireless networking community often represents gains and losses in *decible* (db), defined as

$$G^{db} = 10 \times log_{10} G \tag{2.6}$$

Since the decibel representation only applies to ratios, absolute signal power values are represented relative to a reference value, typically one milliwatt (dbm), so the power budget equation in db becomes

$$p_R^{dbm} = p_T^{dbm} + g_T^{db} + G_R^{db} - L_{FS}^{db} - L_O^{db} \tag{2.7}$$

The link budget captures all the channel effects that can occur when there is only a single path between the transmitter and the receiver. Next, we discuss channel effects in multi-path environments.

## 2.6 The Multi-path Effect

A transmitted signal can reach a receiver in several of ways, including a direct line of sight (LOS) and reflections off large objects such as walls (ray view of a signal). In addition, as shown in Fig. 2.8, the signal can also reach a receiver via diffraction caused by the edge of "large" objects, and by scattering off "small" objects. In this context, object size is relative

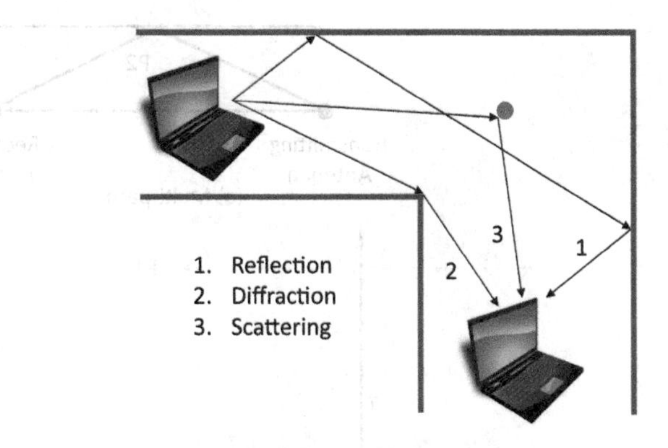


**Fig. 2.8** Signal propagation modes (Laptop figure from https://pixabay.com/static/uploads/photo/2012/04/13/20/24/laptop-33521_640.png. Reused under license https://pixabay.com/service/license/)

to the wavelength of the signal. The receiving antenna will receive a signal that is the sum of all the signals received over these paths.

This *multi-path* effect can be both beneficial and harmful to wireless communication. For example, multi-path can enable communication even when there is no direct LOS path between a sender and receiver. However, it can also result in significant attenuation and distortion of the received signal. In this section we look in more detail at how multi-path impacts wireless communication, focusing first on a signal that has a single frequency, and then on wideband signals that use a wider frequency band.

### 2.6.1 Multi-path Concepts

Figure 2.9a shows a simple example of a transmitter sending a signal to receiver over a LOS path and a reflected path. The receiver gets two copies of the signal, but these copies will arrive with a different signal strength and a small offset in time because the path lengths are different. Path P2 is longer than path P1. Focusing on the case of a single-frequency signal, which we will refer to as a carrier signal for simplicity, we have to consider three scenarios:

- **Constructive Interference:** The two signals arrive in phase, either because the path lengths are the same or they differ by a multiple of a wavelength, i.e., $L2 - L1 = n \times \lambda$, where $\lambda$ is the wavelength of the signal and $n$ is an integer. As shown in Fig. 2.9b, the two signals will arrive in phase at the receiver and they will strengthen each other.
- **Destructive Interference:** The two signal copies arrive 180 degrees out of phase, i.e., $L2 - L1 = (n + 0.5) \times \lambda$. The signals will partially cancel out each other's energy, resulting in a signal strength reduction relative to the strongest signal. If the signals have the same strength, they will cancel each other out as is shown in Fig. 2.9c.
- **The general case**: The two signals reach the destination with random phase and amplitude difference. The result is still a signal with the same frequency, but there are changes in both the phase and offset relative to the two received signal copies.

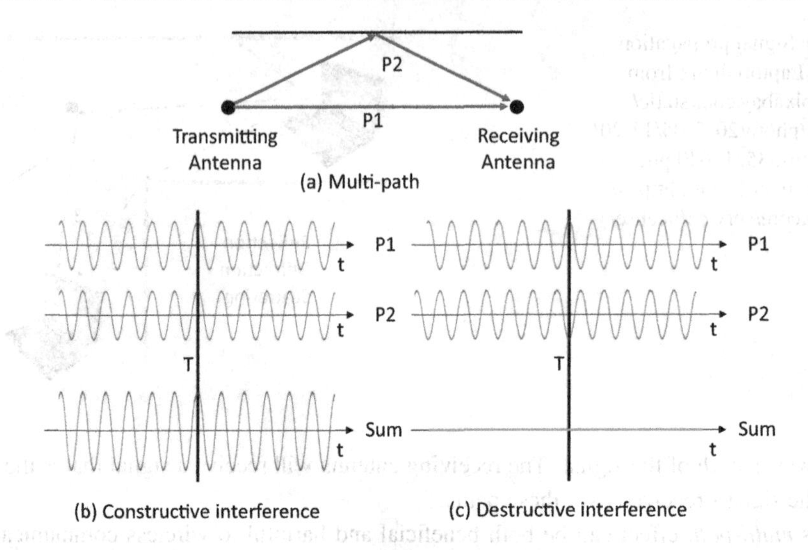


**Fig. 2.9** Two example outcomes of multi-path interference outcomes (Sine wave from https://www.wisc-online.com/assetrepository/viewasset?id=4107 by "unknown". Reused under license CC-BY-NS 2 (https://creativecommons.org/licenses/by-nd/2.0/))

The net outcome is that for a single frequency signal, multipath impacts the amplitude of the signal, resulting a strong signal in the best case to a zero signal in the worst case. This lead to unpredictable network performance for users. Phase shifts may also be a problem if modulation uses the signal's phase (Sect. 2.2).

In real world environments, multi-path effects are much more complex than the simple scenario considered in Fig. 2.9. First, in practice there will often be more than two paths. For a carrier signal, the three outcomes discussed above still apply. Second, the impact of multi-path on the signal depends strongly on the location of the device. Using Fig. 2.9a as an example, if we move either of the two devices, the lengths of the two paths P1 and P2 changes, so the difference in path length, measured in meters and wavelengths, will change as well. As a result, how the signal copies add up will change, impacting link capacity.

This is illustrated in Fig. 2.10, which shows the impact of signal strength variations at a receiver in different locations for a two-path scenario with a stationary transmitter. The z-axis shows the probability of reception, which is strongly correlated with signal strength is a two dimensional area (x and y-axis). For stationary users, this means that there will be "good" and "bad" locations for their wireless device, while mobile users will see changes in network performance as they move around. The results in Fig. 2.10 are for a signal in the 900 MHz band, so the wavelength is 33 cm. We see that even small changes in location of only a few centimeter can already result in big differences in performance. Since multi-path effects are the result of path length differences measured wavelengths, at higher frequencies, performance will change for even small changes in location. For example, WifF uses the 2.4, 5, and 60 GHz bands, where wavelengths are 12.5 cm, 6 cm, and 5mm respectively.

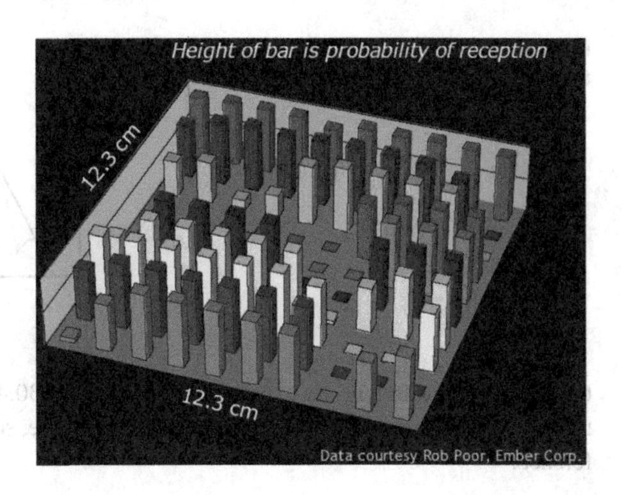


**Fig. 2.10** Multi-path effect depends on location (Figure from slide 15 of https://csis-website-prod.s3.amazonaws.com/s3fs-public/legacy_files/files/attachments/040519_poor.pdf. Reused with permission from Silicon Labs)

### 2.6.2 Multi-path in Practice

So far, we have focuses on single frequency signals but wireless communication uses frequency bands that range from a few 10 s of KHz to 10 s of MHz, so the signal's energy is distributed over a range of frequencies. For each of the frequency components of the signal, the pathlength difference measured in wavelengths will be different, so the impact of multi-path will differ across the frequency components. Specifically, for a path $L$ with length $L_m$ meter, the pathlength $L_w(f)$ in wavelengths at frequency $f$ is

$$L_w(f) = L_m/\lambda(f) \tag{2.8}$$

where $L_\lambda(f)$ the wavelength at frequency $f$.

For narrowband signals, signals for which the width of the band is very small compared with the center frequency, the difference in $L_w(f)$ across the frequency components of the signal is small so the impact of the multi-path effect is very similar across the frequency band, i.e., they experience the same attenuation and phase shifts. In contrast, for wideband signals, some frequencies in the band may experience constructive and others may experience destructive interference. In other words, attenuation of the signal is *frequency selective*. To make this more concrete, let us consider the following example.

**Example 2.2** Let us consider an example of two devices communication using 802.11n in the 5 GHz band with a channel width of 80 MHz. The two edges of the channel of are at frequencies 5 and 5.08 GHz, and the channel between the two devices has two paths. Assume the lowest frequency components (5 GHz, wavelength 6 cm) of the channel experiences constructive path interference since its pathlength difference is 30 wavelengths (1.8 m). The highest frequency component of 5.08 GHz has a wavelength of 5.91 cm, so

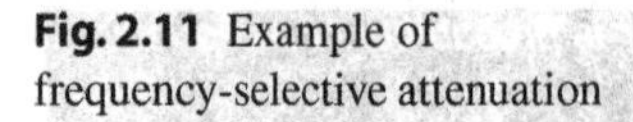

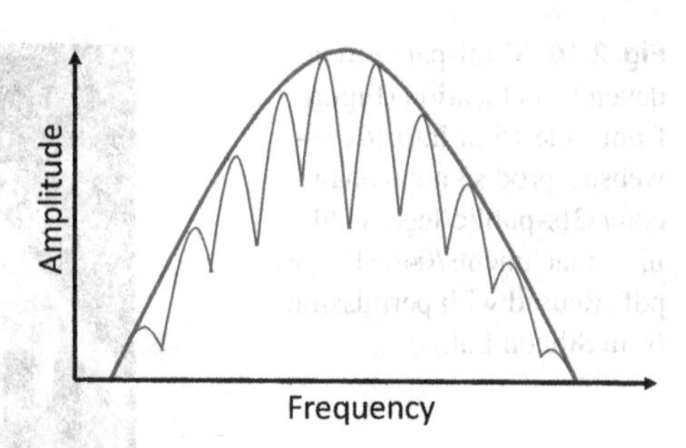


**Fig. 2.11** Example of frequency-selective attenuation

the path length difference it experiences is $\frac{1.8\,mm}{5.91\,cm} = 30.46$ wavelengths. This means that two copies of the signal arrive almost 180° out of phase, so it experiences destructive interference.

The impact of multi-path on a wideband signal is illustrated in Fig. 2.11. The graph shows an example of the amplitude of a received signal as a function of frequency, ignoring noise. The red curve shows received signal strength assuming a channel with a single path, i.e., attenuation is constant across the band. The shape of the received signal is similar to that of the transmitted signal. However, with multi-path (shown in blue), we see much stronger attenuation at specific frequencies that result in strong destructive interference between some paths. Frequency-selective attenuation distorts the signal, which can impact the receiver's ability to demodulate the signal correctly.

### 2.6.3 Inter-Symbol Interference

In the previous section we discussed the impact of the multi-path effect on the wireless signal at the receiver. Multi-path can also have an impact at the level of symbols (Sect. 2.2). Figure 2.12 shows an example of a transmitter sending a stream of symbols, slightly separated in time (top figure). The bottom figure shows the signal received over a channel with three paths. We see that there are three copies of each symbol, with different delays and attenuations. Because the path length differences are high, the copies of consecutive symbols overlap, despite the fact that they are slightly separated in time. For example, parts of the blue symbol overlap with the previous (green) symbol and next (red) symbol. This makes it very difficult to correctly demodulate the signal.

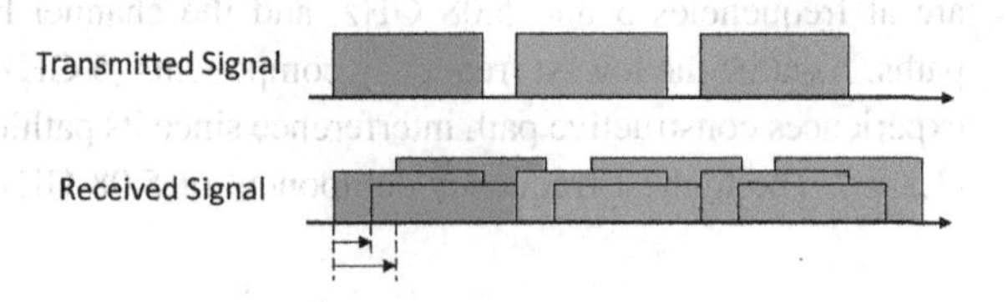


**Fig. 2.12** Example of inter-symbol interference

Inter-symbol interference (ISM) requires pathlength differences that are on the order of a fraction of a symbol time, which are significantly higher than those needed for the multi-path effects discussed in the previous section (order of carrier wavelength). Typical symbol times are on the order of a few μs to tens of μs, resulting in symbol times of 100 m to 10 s of kilometers. Let us look at a simple example:

**Example 2.3** Assume two devices are communicating with a symbol rate of 0.25 MSymbol/sec, a rate used in some WiFi versions, then the symbol time is 4 μs. Assuming symbols are sent back-to-back over a channel with two paths, the path length difference needed for two consecutive symbols to overlap by 10% is

$$0.1 \times 4\,10^{-6} \times 3\,10^{8}\,\text{m} = 120\,\text{m} \tag{2.9}$$

Considering that WiFi is designed for indoor environments, path length differences are typically a small fraction of this. In contrast, LTE is used used outdoors and basestations have a much larger coverage area than Wifi, so path length differences will be much higher. For this reason LTE has a longer symbol time of 66.7 μs. Technologies such as OFDM, described in the Sect. 3.6, use additional mechanisms to further reduce the impact of ISI.

## 2.7 Mobility: Slow and Fast Fading

Wireless channels for which either one, or both, of the communicating devices move, and/or objects in the physical environment move, will experience changes in the path loss. This is called *fading*. There are two types of fading, called *slow fading* and *fast fading* depending on whether the channel is subject to multi-path effects. We discuss both in more detail.

### 2.7.1 Slow Fading

The path loss equation (Eq. 2.3) shows that when there is no multi-path effect, the attenuation can change in a few ways. First, when the transmitter or receiver, or both, are mobile, the distance $d$ is likely to change, and antenna gains $G_t$ and $G_r$ can also change if their relative orientation changes. In addition, when mobile objects fully or partially block the direct LoS path between the two nodes, the path loss coefficient $n$ will change. The net effect is that the attenuation of the channel changes. However, attenuation will change relatively slowly, since it is directly caused by the physical movement. For example, when a person using a mobile device moves away from a basestation at 5 km/h, the distance $d$ between the two devices increases by 1.4 m/s. Unless the person is very close to the AP (e.g., a WiFi basestation in the same room), this is a modest rate of increase in the distance. As a result, the changes in the path loss will be slow, so this effect is called *slow fading*. Of course, in for example vehicular environments, the path loss will change much faster.

### 2.7.2 Fast Fading

In a multi-path environment, we also need to consider frequency-selective attenuation, which is caused by differences in the lengths of the paths. When either the mobile devices or objects in the environment move around, many, if not all, path length will change, so the differences in the path lengths will also change. Changes in the path length difference on the order of a wavelength alters how the signals on the different paths add up (e.g., in a constructive or destructive way). As a result, the attenuation of the signal will change very fast, so this effect is called *fast fading*.

Figure 2.13 illustrates the effects of both slow and fast fading. It shows the signal strength (y-axis) of a received signal of two stationary devices communicating in an environment with people moving around. We see rapid fluctuations of the signal caused by fast fading. The average signal strength over a short time window is relatively stable, but we do see some fluctuation, probably by people partially blocking the line-of-sight path between the two devices.

Figure 2.14 shows a similar result between a mobile and a stationary device. We again see fast fading, but this time the short-term average signal strength changes much more as the mobile device moves around.

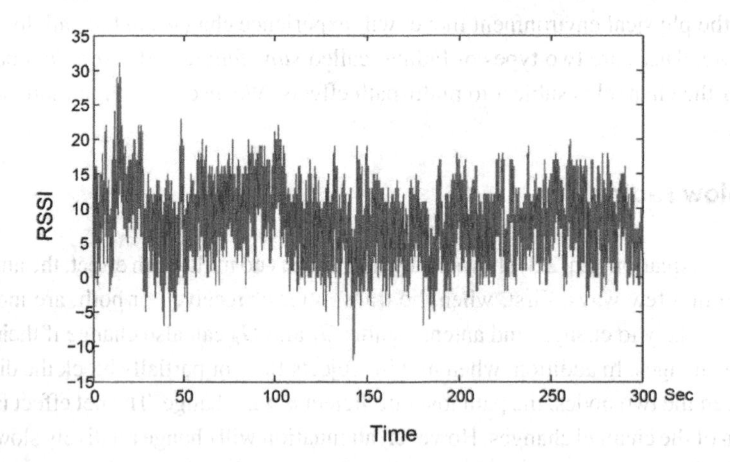


**Fig. 2.13** Fading for a channel between stationary devices (Figure taken from the author's paper Judd et al., 2007 under the ACM's reuse policy https://authors.acm.org/author-resources/author-rights)

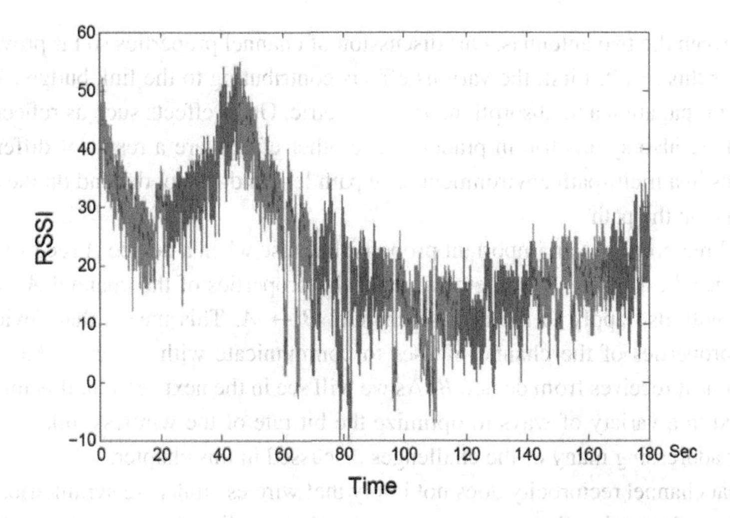


**Fig. 2.14** Fading for a channel between stationary and a mobile device (Figures taken from the author's paper Judd et al., 2007 under the ACM's reuse policy https://authors.acm.org/author-resources/author-rights)

### 2.7.3 Impact of Frequency and Speed on Fading

The effect of fading depends strongly on the wavelength of the signal, since it is determined by the difference in path length measured in number of wavelengths. This means that at higher frequencies (e.g., at 5 GHz, wavelength 6 cm), even small movements by the communicating devices or objects in the environment can easily affect channel conditions, compared with lower frequencies (e.g., 900 MHz, wavelength 33 cm).

Fading is also strongly affected the speed at which the communicating devices or objects in the environment move. For a pedestrian walking at a leisurely speed of 3 $\frac{\text{km}}{\text{h}}$,or 83 $\frac{\text{cm}}{\text{s}}$, the speed with which the channel changes is 2.6 wavelengths per second for a 900 MHz channel. However, a 5 GHz channel will change at a rate of 13.8 wavelengths per second. Achieving good wireless performance at higher speed requires the physical layer to adapt much more quickly to the channel conditions (Camp & Knightly, 2008). Clearly, wireless communication involving cars (e.g., going 90 km/h) or high-speed trains (300 km/h) is very challenging!

## 2.8 Channel Reciprocity

Channel reciprocity means informally that the properties of the wireless channel between the sender and the receiver are the same in both directions, i.e., the channel is symmetric. In this case "channel" includes all the signal propagation effects that impact the signal as it travels between the two antennas. Our discussion of channel properties so far provide some intuition for this result. First, the various effects contributing to the link budget, including free space propagation and absorption, are symmetric. Other effects such as reflections and absorption are also symmetric in practice. The other effects are a result of differences in path lengths in a multi-path environment. The path length does not depend on the direction of the signal on the path.

Channel reciprocity is an important property, because when a device $A$ receives a signal from a device $B$, it can learn something about the properties of the channel $A \rightarrow B$. The properties will also apply to the reverse channel $B \rightarrow A$. This means that device $A$ can learn the properties of the channel it uses to communicate with device $B$ based on the transmissions it receives from device $B$. As we will see in the next section, this information can be used in a variety of ways to optimize the bit rate of the wireless link. In fact, it is critical for addressing many of the challenges discussed in this chapter.

Note that channel reciprocity does not imply that wireless links are symmetric. "link" is datalink layer abstraction that captures properties that are directly relevant to higher levels of the protocol stack such as bit rate and packet delivery rate. The reason is that besides channel properties many other factors impact the performance of the wireless datalink layer. Examples include the transmit power, quality of the radio, radio setting, and the noise level at the receiver. All these factors can be different on the two devices, and they directly impact the bit rate that can be supported.

# Optimizing Throughput at the PHY Layer

# 3

## 3.1 Modulation

We first present general properties of modulation and then introduce Quadrature Amplitude Modulation, a widely used modulation scheme. Next, we discuss how forward error correction can be used to further improve throughput. Finally, we describe dynamic bit rate adaptation, which adapts the modulation bit rate to channel conditions.

### 3.1.1 Tradeoffs in Modulation

Optimizing modulation is the first step in optimizing the performance of a link. The bitrate or Throughput $T$ (in Mbps) of a wireless channel can be represented as:

$$T = S_{SYM} \times R_{SYM} \tag{3.1}$$

were $R_{SYM}$ and $S_{SYM}$ are the symbol rate and symbol size respectively. A symbol represents a number of bits that is treated as a single unit for modulation purposes (Sect. 2.2). The equation suggests that we can potentially increase the link rate almost arbitrarily by increasing the symbol size or symbol rate, or both. Unfortunately, in practice, as we increase the bit rate, signal attenuation and distortion, as described in the previous chapter, will eventually result in the receiver misinterpreting symbols.

Let us first consider the symbol size. Figure 3.1 shows an example of using Amplitude Modulation for symbols representing one and three bits, resulting in symbols with 2 and 8 different values. We see that as we increase the symbol size, the signal strength values representing different symbols move closer together, making it harder for the receiver to distinguish them. As a result, on channels with high channel attenuation, frequency selective fading, and/or noise, such as mobile environments and long links, it will not be practical to



P. Steenkiste, *Introduction to Wireless Networking and Its Impact on Applications*,
Synthesis Lectures on Mobile & Pervasive Computing,
https://doi.org/10.1007/978-3-031-27466-4_3

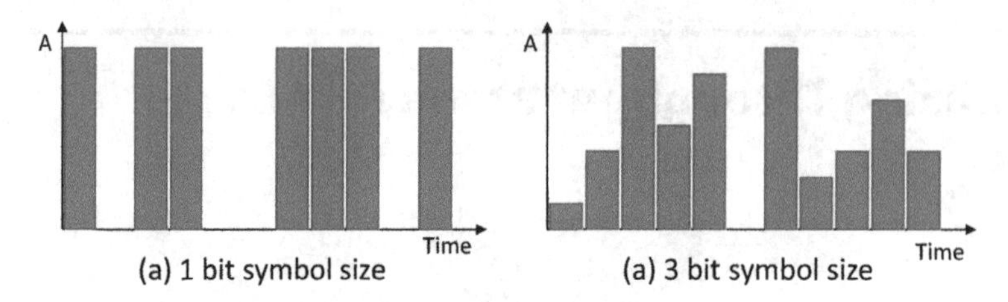


**Fig. 3.1** AM modulation for 2 symbol sizes

use large symbol sizes. However, larger symbol sizes can be used to increase throughput on channels with low attenuation and distortion, for example, a home network with limited mobility.

Second, when we increase the symbol rate, the symbol time, the time it takes to send/receive one symbol, decreases so the receiver has less time to identify the properties of the signal (amplitude in our example), which again increases the chances of bit errors. In addition, shorter symbol times means that the impact of inter-symbol interference can become significant (Sect. 2.6.3). This puts an upper bound on the symbol rate. The effects limiting symbol size and rate, and thus the maximum bit rate a wireless a link can support, depend on the physical environment, and also change over time.

### 3.1.2 Quadrature Amplitude Modulation (QAM)

A sending radio can modulate a carrier signal by change its frequency, amplitude, and/or phase (Sect. 2.2). QAM is a modulation technique that is widely used in high-speed wireless technologies such as WiFi and cellular. Informally, QAM creates symbols by combining amplitude and phase modulation. Figure 3.2a shows an example for a two-bit symbol. The carrier (shown in red in the figure) is aligned along the x-axis, with a certain amplitude and zero phase offset. The four symbols are shown as black dots annotated with the symbol value. In this simple example, the sender only changes the phase of the carrier: the four symbol values correspond to a change of 45°, 135°, 225°, and 315°, for symbol values (1,

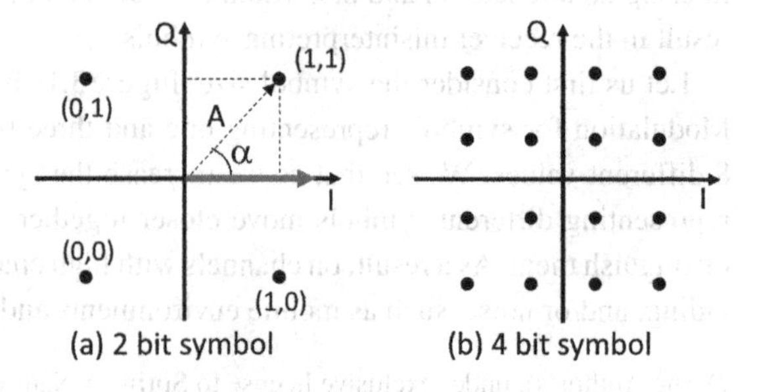


**Fig. 3.2** Two examples of QAM modulation

1), (0, 1), (0, 0), and (1, 0) respectively. Figure 3.2b shows and example of a 4 bit symbol that combines phase and amplitude modulation. This visual representation of the symbols in a 3D plane is referred to as a constellation. In this section we discuss the benefits offered by QAM.

### Flexible Control over the Robustness of Modulation

First, the channel effects described in the previous chapter will change the phase and amplitude of the signal that the receiver has to demodulate. This effectively results in small shifts in the locations of the "dots" representing the symbols in the constellations, as shown in Fig. 3.2. By spreading symbols evenly in the constellation, we can minimize the chance that the receiver will incorrectly map the received signal to the wrong symbol. Of course, the more challenging the channel, the larger the shift, so under very challenging conditions, the receiver can still misinterpret the symbol as is illustrated in Fig. 3.3. *Constellation S* represents the signal transmitted by a sender. *Constellations* R1, R2, and R3 show the constellation at three receivers with different levels of background noise. The red dots show the signal that was transmitted while the small blue dots show the received signal for a large number of symbols. We see that as the noise level increases, it will become harder for the receiver to correctly identify the transmitted symbol value.

### Easy to Generate

Second, QAM signals are fairly easy to generate. Specifically, they are the sum of two modulated carriers that are offset in phase by 90°:

$$s(t) = I(t) \times sin(2\pi \times f_c \times t) + Q(t) \times sin\left(2\pi \times f_c \times t + \frac{\pi}{2}\right) \quad (3.2)$$

where $s(t)$ is the modulated signal that is transmitted, $f_c$ is the carrier signal, and $I(t)$ and $Q(t)$ are called the in-phase and quadrature components of the transmitted signal; they are effectively the amplitudes on the x and y-axis for the symbol in the constellation. The receiver obtains (approximations of) the $I$ and $Q$ values by demodulating the received signal using the in-phase (for $I$) and 90° out of phase carrier (for $Q$). Note that a radio can use the same technique for many constellations representing symbols of different sizes.

### Numeric Representation of the Transmission Process

Third, QAM modulation enables modeling the transmission process by representing the transmitted and received signals complex numbers. For example, the transmitted signal can be viewed as a complex number $s_T$, with $I_T$ and $Q_T$ as the real and imaginary parts. As shown in Fig. 3.4a, this numeric representation is equivalent to the amplitude $A$ used to modulate. Similarly, the changes in amplitude and phase changes inflicted by the channel on the signal as it travels between the transmitting and receiving radios can be viewed as a vector in the

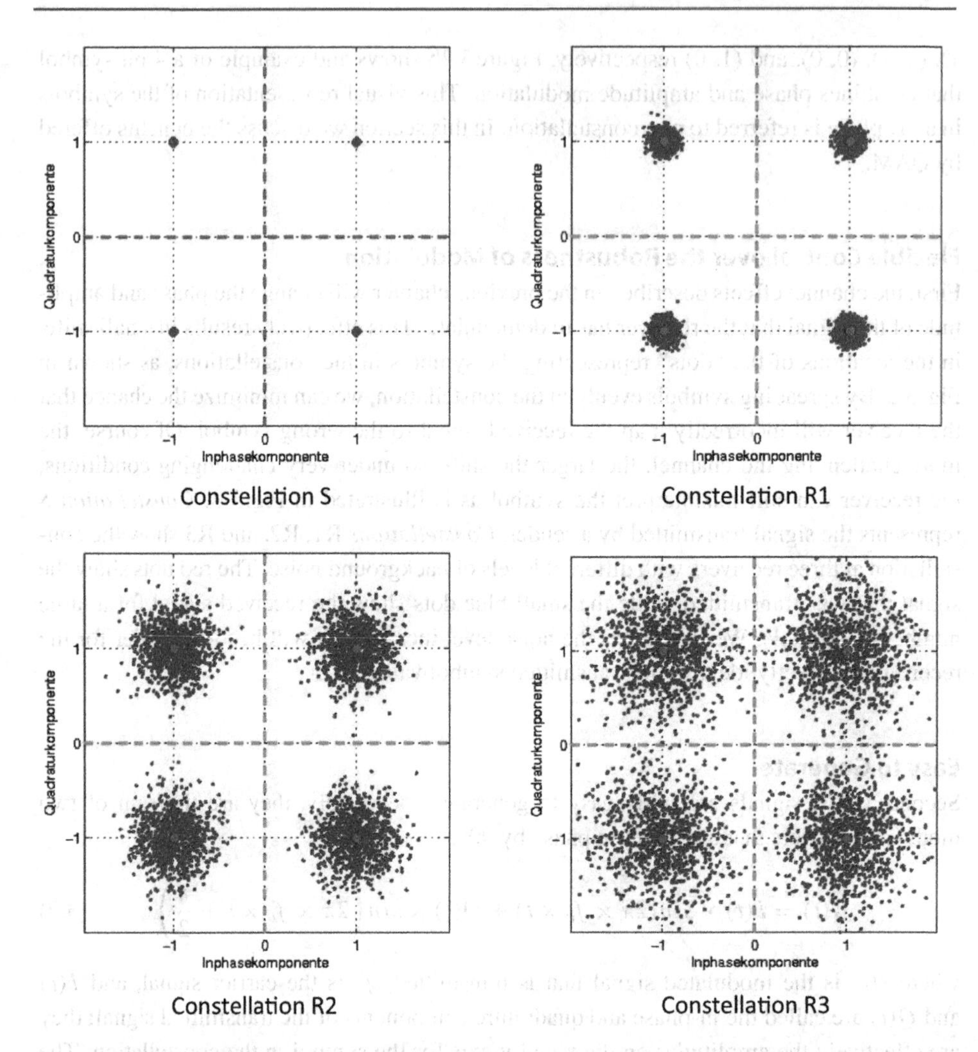


**Fig. 3.3** Impact of noise on a QAM modulated signal (Four constellation figures reused from https://commons.wikimedia.org/wiki/File:4qam_constellation_new.png, https://commons.wikimedia.org/wiki/File:4qam_constellation_noisy_sigma001.png, https://commons.wikimedia.org/wiki/File:4qam_constellation_noisy_sigma01.png, and https://commons.wikimedia.org/wiki/File:4qam_constellation_noisy_sigma025.png by Stickle (https://de.wikipedia.org/wiki/Benutzer:Sticle) Reused under under CC BY-NS 2 https://creativecommons.org/licenses/by-nd/2.0/)

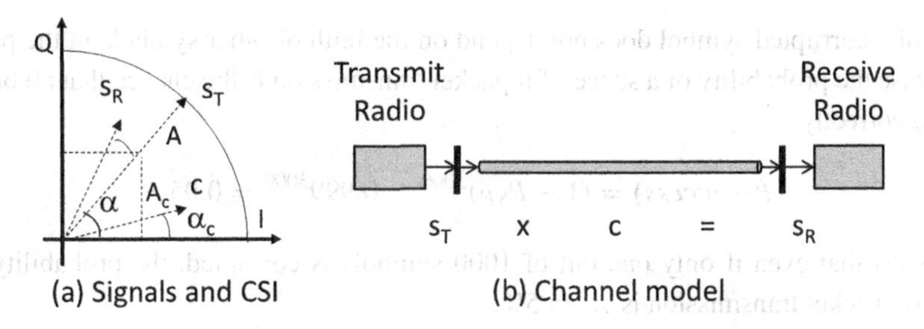

(a) Signals and CSI (b) Channel model

**Fig. 3.4** Representation of wireless transmission using CSI

constellation with an amplitude $A_c$ and phase $\lambda_c$, representing the change in amplitude and phase. It can also be represented as a complex number $c$, which is called the *Channel State Information* (CSI). The CSI changes over time as a result of fading, so we will represent it as $c(t)$. The CSI is also a function of the frequency $f$ because of frequency selective fading (see Sect. 2.6.1), but for simplicity, we will not explicitly show its dependence on $f$.

Figure 3.4b shows how we can now mathematically represent the wireless transmission process for a packet. The CSI $c(t)$ summarizes the impact of both the antennas and the wireless channels on the wireless signal $s_T(t)$ sent by the transmitting radio. The signal available to the receiving radio $s_R$ is the product of $s_T$ and $c(t)$ (Fig. 3.4b):

$$s_R(t) = c(t) \times s_T(t) \tag{3.3}$$

The multiplication of the two complex numbers results in a value $s_R$ with an amplitude equal to product of the amplitudes $A_c$ and $A_T$ and a phase that is the sum of the two phases $\lambda_T$ and $\lambda_c$, which correctly captures the impact of the antennas and channel on the transmitted signal.

As we will see later in the chapter, the simple numeric representation of both QAM modulated signals and the CSI allows wireless radios to aggressively optimize wireless communication.

## 3.2 Combining Modulation and Forward Error Correction

While aggressive modulation is a powerful way to increase channel throughput, it comes with the risk that many packets will be received with errors. Let us use a simple example to illustrate this:

**Example 3.1** Assume a device is sending a 1500 Byte packet to a receiver using QAM modulation with four-bit symbols Since each byte requires two symbols, the packet will be sent as 3000 symbols. Let us assume that the probability that the receiver misinterprets a symbol is $P_{SE} = 0.1\%$. For simplicity. we will assume that errors are independent, i.e., the chance of a corrupted symbol does not depend on the faith of other symbols in the packet. In that case the probability of a successful packet transmission is the chance that all bits are received correctly:

$$P(success) = (1 - P_{SE})^{3000} = 0.999^{3000} = 0.05 \tag{3.4}$$

This means that even if only one out of 1000 symbols is corrupted, the probability of a successful packet transmission is about 5%.

As we discuss in more detail in the next chapter, datalink protocols use error detection so the receiver can verify that the received packet is correct, i.e., there are no bit errors. Packets with errors are discarded and may be retransmitted by the sender depending on the protocol. Retransmissions are expensive: not only is the packet delayed, but they waste network capacity that could have been used for new packets. Wired networks have extremely low bit (symbol) error rates, so this is not a concern. However, wireless networks have higher error rates, so the likelihood of a packet corruption can be significant, as the above example illustrates. One way of reducing the symbol error rate is to reduce the symbol size, as discussed in the next section. Unfortunately, picking conservative symbol sizes reduces the bandwidth of the link, directly impacting both user performance and network capacity.

An alternative is to use Forward Error Correction (FEC) which relies on the use of error-correcting codes that allow the receiver to recover from certain symbol errors without requiring retransmission. It is a proactive technique, while retransmission is a reactive technique. Not surprisingly, the cost of using error-correction, compared with error-detection, is higher, i.e., $S_{ECC} > S_{EDC}$, where $S_{ECC}$ and $S_{EDC}$ are the sizes of the error-correcting and detecting codes that are added to the packet. However, when FEC is used, modulation can use larger symbols, so there is a tradeoff between the cost of using FEC and the increase in bit rate and link bandwidth it allows. Technologies such as WiFi and cellular use both error correction and retransmission to recover from errors.

## 3.3 Variable Bit Rate Protocols

Mobile wireless devices are used in diverse environments that can have widely different channel conditions, e.g., home office, crowded coffee shop, and outdoors. Even in a single location, channel conditions often change as a result of changes in the physical environment (opening or closing doors) and mobility. Clearly, having wireless devices use a single bit rate for transmitting packets will be very inefficient, since this fixed rate will either be too aggressive or too conservative in most environments

For this reason, modern wireless protocols support multiple transmission rates. Based on the discussion in the previous section, radios can change their transmission rate in two ways. First, they can pick different symbol sizes and rates. For simplicity, protocols typically use a fixed symbol rate and they only change the symbol size (the number of bits per symbol).

**Table 3.1** Modulation and bit rates for 802.11a

| Mode | Modulation | Symbol size | Coding rate | Bit rate (Mbit/s) |
|---|---|---|---|---|
| 1 | BPSK | 1 | 1/2 | 6 |
| 2 | BPSK | 1 | 3.4 | 9 |
| 3 | QPSK | 2 | 1/2 | 12 |
| 4 | QPSK | 2 | 3/4 | 18 |
| 5 | 16-QAM | 4 | 1/2 | 24 |
| 6 | 16-QAM | 4 | 3/4 | 36 |
| 7 | 64-QAM | 8 | 1/2 | 48 |
| 8 | 64-QAM | 8 | 3/4 | 54 |

Second, they can use different levels of FEC by changing how many bits are used for error correction purposes.

In practice, radios cannot pick arbitrary values for the symbol size and FEC level. The reason is that for every packet transmission the transmit and receive radios, which may be built by different companies, need to agree on the symbol size and FEC level that is used. For this reason, wireless protocols define a set of bit rates that devices must support. Each bit rate has an identifier that the sending radio uses to inform the receiving radio what symbol size and error correction level is used for a packet. This information is included in a header.

As an example, Table 3.1 shows the bit rates for the different "modes" supported in 802.11a. The table also shows the modulation type, the symbol size, and the coding rate for each bit rate. The constellation diagrams in Fig. 3.2(a) and (b) are for QPSK and 16-QAM modulation. The coding rate $u/t$ represents the number of useful data bits $u$ per bits transmitted $t$, so a rate of $1/2$ means that half the bits sent are used for error detection and correction. All bit rates use the same symbol rate, so the bit rate is fully determined by the modulation type and coding rate. Later versions of WiFi further increased the bit rate by using larger symbols, in addition to other techniques for improving throughput, as described later in this chapter.

## 3.4 Bit Rate Adaptation

For wireless technologies that support multiple bit rates, wireless devices have to pick the best bit rate, given the channel conditions. While the protocol standards specify a set of supported bit rates and protocol features that allow wireless transmitters and receivers to use these bit rates, the algorithm that is used to pick a rate is typically left to the device manufacturer.

Bit rate algorithms have become more sophisticated over time. Using WiFi as an example, early WiFi algorithms were based on trial and error. For each destination, a WiFi device would

**Fig. 3.5** Bit rate selection based on path loss

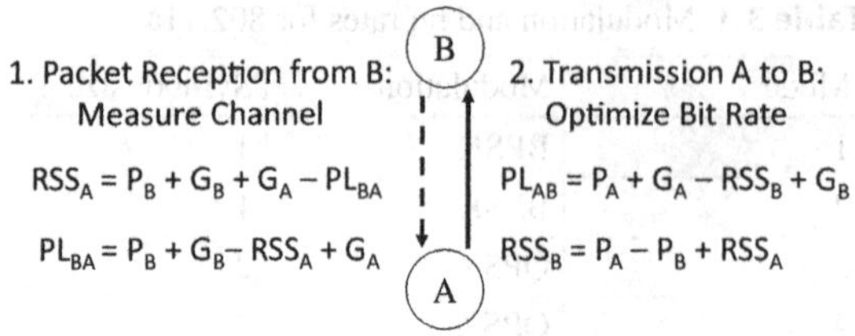

start transmitting at a certain bit rate. After a certain number of successful transmissions, it would increase the bit rate opportunistically. Similarly, after one or more losses, the rate would be decreased. Many algorithms have been proposed that differ in the logic used to decide when to increase/decrease rates and by how much.

The next generation of algorithms used channel state information to decide what bit rate to use. The key idea is that based on the channel reciprocity property (Sect. 2.8), the sender can estimate the channel properties based on the packets it receives from the intended receiver. Early versions simply used the path loss of the channel as the channel state, as illustrated by the Charm (Judd et al., 2007) protocol. Figure 3.5 shows two devices A and B exchanging packets. A wants to optimize its packets transmissions to device (solid arrow) by picking the best bit rate, based on an estimate of the received signal strength $RSS_B$.

The figure shows the equations for the received signal strengths at A and B, $RSS_A$ and $RSS_B$, and for the pathloss in both directions based on the transmit powers $P_A$ and $P_B$, and the antenna gains $G_A$ and $G_B$. Channel reciprocity states that $PL_{AB} = PL_{BA}$, so

$$P_A + G_A - RSS_B + G_B = P_B + G_B - RSS_A + G_A \quad (3.5)$$

or

$$RSS_B = (P_B - RSS_A) - P_A \quad (3.6)$$

Device A knows $P_A$ and can measure $RSS_A$, while $P_B$ can be provided by device B. The Charm protocol was implemented for 802.11 and it used a look up table to map the $RSS_B$ estimate onto the best bit rate. As we discus later in the chapter, today's wireless networks use much more sophisticated technologies which require access to more fine-grain channel state information, but the optimization of throughput is based on the same idea.

## 3.5 Diversity in Space, Time, and Frequency

Efficient modulation is an important tool in optimizing the bit rate of a wireless link and it directly addresses many of the wireless challenges identified in the previous chapter. Challenges generally fall in one of three categories:

- **Space**: Due to multi-path, the quality of the signal at the receiver can be radically different in locations that are separated by a distance of half a wavelength or less (Sect. 2.6.1). This can lead to very unpredictable wireless performance since the link rate strongly depends on the exact location of the device.
- **Frequency**: In a multi-path environment the signal suffers from frequency-selective attenuation (Fig. 2.11). This distorts the signal, affecting user performance. Frequency selective attenuation is a major concern for technologies that use wide channels, e.g., WiFi and cellular.
- **Time**: Due to mobility, the quality of the signal changes over time when one or both of the devices, or any other objects in the environment, move, resulting in slow fading (Sect. 2.7). In a multi-path environment, it also leads to frequency-selective fast fading where the frequency ranges that suffer from high attenuation change over time. Both fast and slow fading lead to variable performance for users.

Advanced modulation techniques help in optimizing bit rate in challenging conditions, but it mostly deals with general distortions of the signals (shifts in the phase of amplitude) and it does not addresses the above challenges. In the next few sections, we introduce techniques that leverage *space, frequency and time diversity* to address these challenges.

## 3.6 Orthogonal Frequency Division Multiplexing (OFDM)

Orthogonal Frequency Division Multiplexing (OFDM) is designed specifically to address the challenges associated with multi-path environments. This is very important for high-performance wireless technologies because they are very aggressive in optimizing the link bandwidth. First, they use more aggressive modulation, i.e., higher symbol rates and/or larger symbol sizes. Unfortunately, increasing the symbol rate shortens the symbol time and inter-symbol interferences (ISI) becomes a concern in many more environments. Second, they use wider frequency bands. For example, while the early WiFi versions used 20 MHz, 802.11ax can use up to 160 MHz of spectrum. This means that frequency-selective attenuation and fading becomes a much more significant problem.

### 3.6.1 OFDM Concept

The key idea of OFDM is to transmit the data not by modulating a single carrier, but by using a large number of carriers, called subcarriers, each of which uses only a small part of the available frequency band (Halperin et al., 2010). This concept is illustrated in Fig. 3.6. Figure 3.6a shows a signal using a wide frequency band (y-axis) over time (x-axis) using single-carrier modulation. The symbols are represented by colored lines. Since symbols are sent sequentially, they use the full frequency band. This means that (1) the signal is subject

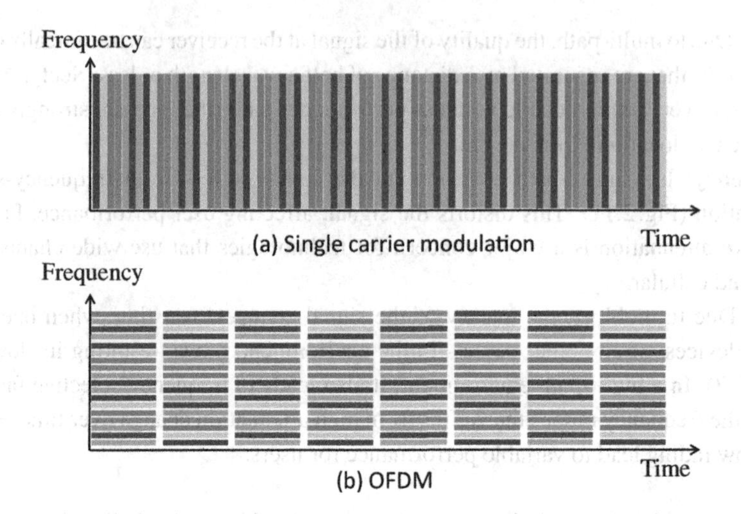


**Fig. 3.6** OFDM's use of subcarriers

to frequency selective fading and (2) the symbol times are very short in the time domain so they are very susceptible to inter-symbol interference.

Figure 3.6b shows how OFDM instead modulates many subcarriers in parallel and each subcarrier only uses a small part of the frequency band. As a result, each modulated subcarrier is a narrow band signal, so frequency-selective fading is not a concern. In addition, since the bit rates supported by the subcarriers are much lower than those of the single-carrier signal in Fig. 3.6a, the symbol times are much longer, making it easier to address ISI. Informally and ignoring many details, when OFDM uses $N$ subcarriers in a spectrum band with bandwidth $B$ Hz, each subcarrier will be a narrow-band signal of $\frac{B}{N}$ Hz, and the symbol times will be roughly a factor $N$ longer.

### 3.6.2 OFDM Optimizations

OFDM makes it possible to directly address several problems caused by frequency selection attenuation. First, subcarriers that are affected by frequency-selective attenuation can be amplified by the receiver. This also amplifies the noise, but the performance penalty this incurs is much smaller than that of a distorted wideband signal. The receiver can also easily adapt when the attenuated subcarriers change over time due to fast fading. In addition, while the longer symbol times reduce the impact of inter-symbol interference (ISI), OFDM also inserts small delays between successive symbols to make sure that copies of successive symbols on different paths do not overlap.

Finally, OFDM directly supports both frequency and time diversity (Sect. 3.5). Blocks of data are sent as a sequence of symbols that use different subcarriers and time slots. Time and frequency dependent fading will generally impact only some of the symbols of each block of data and FEC can be used to recover the data.

### 3.6.3 OFDM Implementation and Use

While OFDM has many benefits, it does require simultaneously transmitting data over subcarriers that use very narrow spectrum bands. In order to avoid interference between adjacent subcarriers, this normally requires the use of narrow guard bands that separate the subcarriers, as is shown in Fig. 3.7a. Unfortunately, with a large number of subcarriers, the the guard bands would waste a fair bit of spectrum, reducing the link capacity.

To avoid this, OFDM uses *orthogonal* subcarriers that can be very tightly packed in the spectrum band, as is illustrated in Fig. 3.7b. Since the orthogonal subcarriers are tightly packed without guard bands, more subcarriers can be transmitted in a given frequency band. The dense packing of orthogonal subcarriers is possible because of the profile the subcarriers in the frequency domain. Each subcarrier has a main "peak" of energy (shown in the figure) with increasingly weaker mirror images on both sides (not shown). The points between adjacent mirror images of the same subcarriers are "nulls" with zero energy. OFDM aligns the main peak of each subcarrier with the "zeros" of the other subcarriers, thus minimizing interference. Clearly this requires that the subcarriers have very precise, equally spaced frequencies. Modulating each subcarrier separately and then placing them at precise frequencies in the target spectrum would be very challenging. In reality, the subcarrier generation and modulation is done entirely digitally in software, which results in a digital signal that is then used to modulate a single carrier.

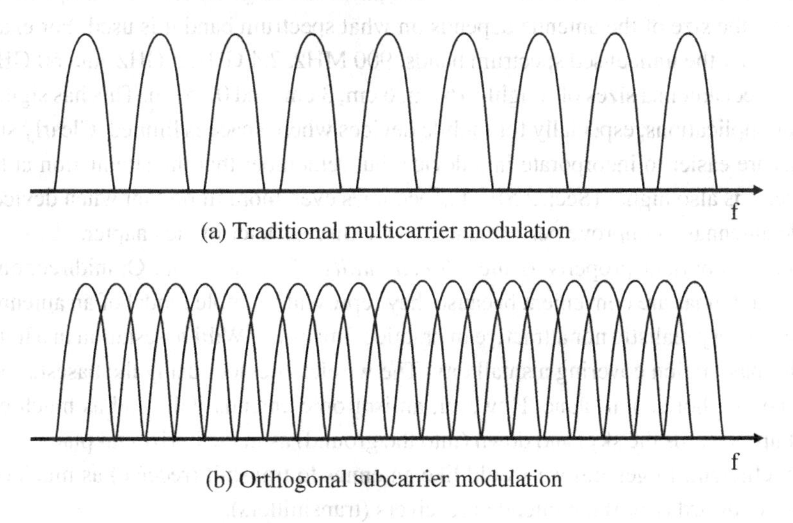


**Fig. 3.7** Tight packing of OFDM subcarriers

OFDM is widely used in WiFi and cellular technologies. As an example, 802.11a supports bandwidths of 20 MHz, 40 MHz, 80 MHz, and 160 MHz, that respectively use 64, 128, 256 and 512 subcarriers, so the spacing between subcarriers is roughly 312.5 KHz. LTE uses a much lower subcarrier spacing of 15 KHz. The reason is that the narrower subchannels have less capacity, which results in longer symbol times which helps LTE address inter-symbol interference. Cellular technologies have much longer transmission ranges than WiFi, so inter-symbol interference is a much more significant problem. since the differences in path lengths are much higher.

## 3.7 Antennas

In this section we first discuss why the choice of antenna is important. We then discuss how multiple antennas can be used to achieve space diversity, resulting in potentially significant performance gains.

### 3.7.1 Performance Considerations

The most important property of its antenna is its *efficiency*. Specifically, when the radio provides a signal to the antenna, we want the antenna to radiate a lot of that energy toward potential receivers. Similarly, we want the receiving antennas to capture as much of the radiated energy as possible, so they can make it available to the receiving radio. It turns out that the efficiency of the antenna depends strongly on the size of the antenna. Efficient antennas are on the order of a half a wavelength of the targeted (carrier) frequency. This means that the size of the antenna depends on what spectrum band it is used. For example, if we consider the unlicensed spectrum bands, 900 MHz, 2.4 GHz, 5 GHz, and 60 GHz, we should expect antenna sizes of roughly 16 cm, 6 cm, 3 cm, and 0.25 cm. This has significant practical implications, especially for mobile devices where space is limited. Clearly smaller antennas are easier to incorporate in a device, but remember that the attenuation at higher frequencies is also higher (Sect. 2.5)! This becomes even more important when devices use multiple antennas to improve performance, as we discuss later in the chapter.

Another important property is the *directionality* of the antenna. Omnidirectional, or *isotropic*, antennas are convenient because they represent a simple model of an antenna, but they are not very realistic nor attractive in practice. Imagine a WiFi basestation in a home, or a cellular basestation covering a small city. The wireless devices using the basestation will mostly be in a horizontal plane. However, an isotropic antenna will send as much energy straight up (towards the sky) and down (into the ground) as in the horizontal plane, which is not very efficient. In general we would like antennas to transmit (receive) as much energy as possible toward (from) the intended receivers (transmitters).

As an example, a widely used antenna for WiFi is called a *dipole* antenna. It is a simple straight conductor. When place in the vertical position, it has the strongest signal in the horizontal plane, which is where the WiFi devices are typically found. The signal strength decreases as the angle of the path relative to the horizontal plane increases and it is zero in the vertical direction.

In some scenarios, devices may want to "point" their antenna to transmit to, or receive from, a specific area or device. For example, a device may want to increase its transmit or receive range towards a device, or reduce interference from other nearby wireless devices. This can be done by using *directional antennas* at the sender and/or the receiver. There are two broad classes of directional antennas. A first class consists of antennas that have a shape that is designed to direct the signal energy to, or receive from, a particular direction. This type of directional antennas is mostly used in more specialized domains, for example for satellite communication or point-to-point long-distance wireless links. A second class of directional antennas uses an array of antennas to perform *beam forming*, as we discuss in Sect. 3.8 below.

The impact of using antennas that optimize communication in specific directions is captured by the transmit and receive antenna gains $G_T$ and $G_R$ in Eq. 2.3. The antenna gain represents the signal strength relative to an isotropic antenna for a particular path. For example, for a *bipole* antenna, paths that are in or close to the horizontal plane will have an antenna gain that is higher than one, while it will be lower than one for paths that are angled up or down.

### 3.7.2 Spatial Diversity: More Antennas is Better!

As we discussed in Sect. 2.6.1, in multi-path environments the signal strength at the receiver depends strongly on the location of the transmit and receive antennas (Fig. 2.10), since differences in path length can result in either constructive or destructive interference between the multiple copies of the received signal. A simple, albeit somewhat brute force, solution to reduce the impact of destructive interference is to connect multiple antennas to the radio. When a receiver, for example, can choose one of two antennas, chances are that at least one of the antennas has a good signal. Clearly this will not work if antennas are too close together, but Fig. 2.10 suggests that if the distance between antennas is at least half a wavelength, they will be impacted differently by the multi-path effect.

More generally, if antennas are separated by at least a half wavelength, their instantaneous channel properties will be uncorrelated. Note that they will share some more general properties (such as attenuation due to free space loss and objects). The half wavelength rule has practical implications. For small devices, such as cell phones, there is clearly a limit to how many antennas can be used, especially at lower frequencies. Laptops and stationary devices such as access points have more flexibility.

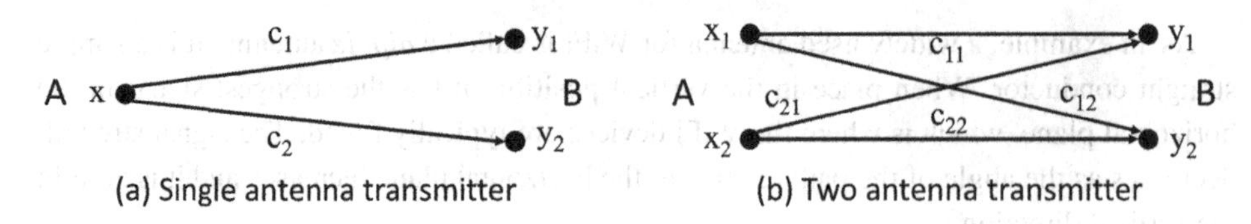


**Fig. 3.8** Selection diversity

Many variants of space diversity have been used over the years. Early systems used *selection diversity*. Figure 3.8a shows an example with a transmitter A with one antenna transmitting to a receiver B with two antennas. $x$ and $y$ represent the transmitted and received signals, $c$ represents channel state information (CSI), and the subscripts identify the antennas that are used on the devices. In this simple case, the receiver B can simply select the antenna with the strongest signal. In general, both devices have multiple antennas, as shown in Fig. 3.8b. In this case, the two devices have to jointly pick which of the four available channels has the best throughput. For example, the A can send a packet on each of its antennas. This allows B to measure the signal strength for all four channels, allowing it to pick the best channel. It can then tell A what antenna it should use. Because of channel reciprocity, the same channel should be used in both directions. Since many wireless technologies use acknowledgements, i.e., sending data involves two-way communication, both nodes continuously receive up to data information about the channel CSI, so they can adapt to changing channel conditions.

While selection diversity can significantly improve performance, it is possible to do much better. The intuition is that two signals generally contain more information than one. To illustrate how we can improve performance using two signals, let us focus on a transmission from node A to node B in Fig. 3.8a. Note that B gets two copies of the signal and instead of throwing away one of the copies, it might for example be better to add the two signals to improve signal strength, resulting in the following SNR for the combined channel:

$$S_y = S_{y1} + S_{y2} + 2 \times N_B \tag{3.7}$$

where $N_B$ is the noise level at node B. Unfortunately, this approach has two drawbacks. First, since we are doubling the noise, adding the signals is unlikely to improve the SNR. Second, the two copies of the signal received by at B's antennas may be out of phase so destructive interference could in fact reduce the signal strength.

The best solution for this example is to use a technique called *maximum ratio combining*: B uses a weighted average of the two signals in a way that (1) gives more weight to the stronger signal and (2) aligns the phases of the two signals so they amplify each other. More precisely, we can represent the transmission using the following equation:

$$Y = C \times x + N \tag{3.8}$$

where $x$ represents the transmitted signal (a complex number), and $C$, $N$, and $Y$ are column vectors representing the channel state, noise levels and received signals at B's two antennas. With maximum ratio combining, B calculates a weighted average for $y_1$ and $y_2$ by multiplying $Y$ with the complex conjugate of $C$, represented as a row vector $C^*$. This gives the most weight to the strongest signal while also adjusting for any phase differences (Shrivastava et al., 2008):

$$Y_{opt} = C^* \times Y = H^*(C \times x + N) \tag{3.9}$$

This optimization requires channel state information (CSI) but it is otherwise easy to implement. In the more general case where both nodes have multiple antennas. The CSI of all the channels is then represented as a matrix, and the sender needs to do some preprocessing of the signals, in addition to the processing at the receiver.

## 3.8 Beam Forming

Another way to use multiple antennas to increase wireless link bandwidth is beam forming. One can think of beamforming as using multi-path to our advantage. Figure 3.9 shows an example of a device A with four antennas using beam forming to optimize communication with node B. The key idea is that A transmits the same signal on each of the four antenna elements but with a different phase. The phase difference across the antennas is tuned so the four signals will combine constructively at receiver B, amplifying the signal. Alternatively, device A can use multiple antennas to avoid interference at node C by adjusting phases so the nodes combine destructively at node C. Increasing the number of antennas allows for more fine grain control over beam forming towards possibly multiple devices. Techniques that use channel state information, similar to those described in the previous section, can be used to control the antennas at the sender and/or receiver to optimize link bandwidth. An important advantage of beam forming is that it allows nodes to easily change the direction that offers high gain.

Beam forming is support on recent versions of WiFi and it is very important for mmWave communication as discussed in Sect. 4.3.6.

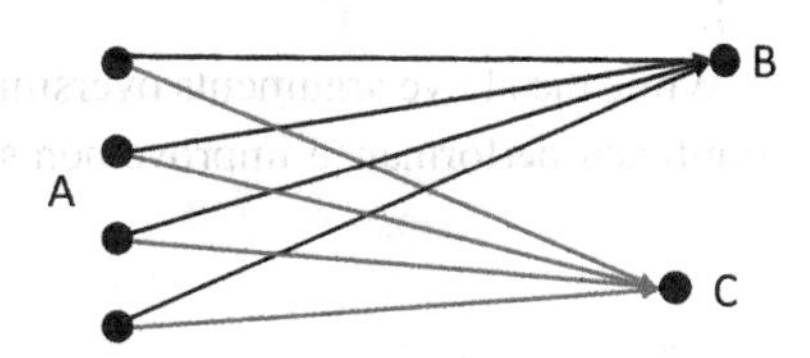


**Fig. 3.9** Example of beam forming

## 3.9 MIMO and Multi-user MIMO

In the previous section we use multiple antennas to improve throughput, but we only send a single data stream and the communicating nodes only use one radio. We now look at how we can further improve performance, for example, if our bandwidth requirements exceed the Shannon limit of single channel.

### 3.9.1 MIMO

Let us first look at wired communication for inspiration. Consider an example of two nodes that need to communicate via ethernet at a throughput of up to 40 Gbps but the maximum ethernet link speed is 10 Gbps. Figure 3.10a shows how we can easily do this by connecting the two nodes with 4 ethernet links that each carry one fourth of the data. Note that since each ethernet carries different data, both nodes will use four independent radios.

As a strawman, we show an equivalent configuration for wireless in Fig. 3.10b. Each node has four radios, each with its own antenna. They send four parallel data streams between transmit/receive radio pairs 4 in the same frequency band. This simple solution will of course not work. In the wired case, the four signals travel in a wire that isolates the four signals from each other, so each receiving radio receives exactly one signal, which it can demodulate and decode without a problem. In the wireless scenario however, each of the four receiving radios will receive copies of all four signals combined, one signal of interest and three strongly interfering signals.

Fortunately, the situation is not as dire as it looks. Let us first provide some intuition for why this might be possible. First, radio 1 at the receiver must retrieve signal 1 sent by the transmitter radio 1, despite the interference. One idea is that radio 1 can use the signals received by radios 2–4 on the receiver to cancel out the interference caused by signals 2–4. This is a type of *interference cancelation*, not unlike how a noise canceling headsets work. Another way of looking at the problem is that each of the four receiving radios receives some information about each of the 4 signals. The question is whether we can combine the information embedded in the four signals to generate good copies of transmitted signals 1–4.

While the above arguments oversimply the problem, it turns out that it is possible to get significant performance improvements by sending parallel data streams. The technique is

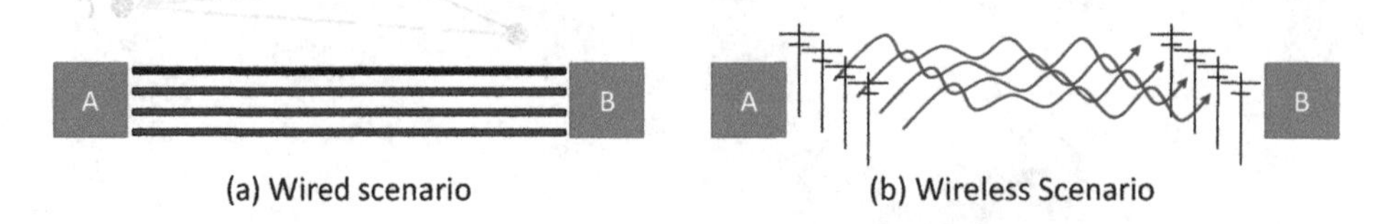


**Fig. 3.10** Sending multiple data streams

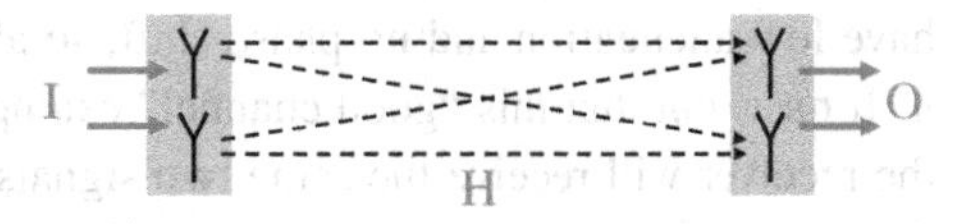


**Fig. 3.11** 2 by 2 MIMO example

called MIMO, which stands for Multiple In Multiple Out (Halperin et al., 2010). Figure 3.11 shows an example of a sender and receiver, each with two antennas and two radios. The goal is to send two parallel data streams. In the figure, capital letters represent column vectors or matrices. Specifically, $I$ is a column vector representing two signals that are modulated with the two data streams by the transmitter, $H$ is the channel state matrix representing the CSI of the four channels, and $O$ represents the two signals that the receiver must demodulate. $O$ can be expressed as a function of $I$ as:

$$O = H \times I + N \tag{3.10}$$

where N represents the noise at the receiver's antennas. In order to successfully demodulate the input signals, the receiver needs access to a copy $O \approx I$, which this simple design cannot provide.

The key idea enabling MIMO is similar to that used to leverage spatial diversity (Fig. 3.8b). In fact, Eq. 3.10 is similar to Eq. 3.8. The only difference is that with MIMO, we are transmitting two different signals representing two data streams, while only a single data stream was sent in the diversity design. Figure 3.12 shows how MIMO can be supported by adding pre and post-processing of the signal in the transmit and receive radios. Specifically, pre and post-processing consists of multiplying the input and received signals $I$ and $C$ with the matrixes $P_T$ and $P_R$ as follows

$$O = P_R \times H \times P_T \times I + P_R \times N \tag{3.11}$$

$P_T$ and $P_R$ are calculated based on the singular value decomposition (SVD) of the channel state matrix $H$. Specifically, $H$ is decomposed as $H = U \times S \times V$, where $S$ is a diagonal matrix and, informally, the inverse of $U$ and $V$ are used as $P_R$ and $P_T$.

In practice, the result $O$ will differ from $I$ because of errors in $H$ and the effect of noise. SVD is computationally inexpensive for small matrices, but note that when we use OFDM for wide-band channels, so we need to measure $H$ and do the computation for all OFDM subcarriers. Finally, because of fading $H$, $P_T$, and $P_R$ need to be updated regularly.

Unfortunately, MIMO cannot always achieve significant performance benefits. Let us use a simple example to illustrate this. Assume that in the above example, all four channels

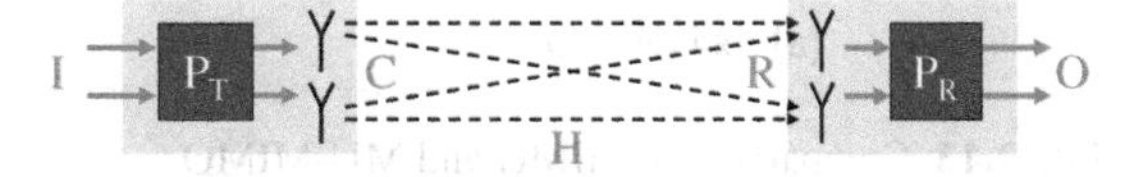


**Fig. 3.12** 2 by 2 MIMO design

have low attenuation and no phase shift, so all four values in $H$ have the same real value $h$. It turns out that this "good channel" example is a worst case scenario for MIMO since the receiver will receive the same two signals on each antenna, i.e., $c_1 = c_2$,. This means that having two antennas does not provide more information than one antenna. In practice, MIMO is most effective when the channels between the transmit and receive antennas are uncorrelated, i.e., they have different properties. One example is non-line of sight scenarios, where the channels experience different multi-path effects.

### 3.9.2 MU-MIMO

MIMO is a powerful way of increasing the bandwidth between two devices, but in order to be effective, a number of application and system level constraints must be met. First, MIMO is mostly useful for large data transfers between two nodes and the benefits will be limited for short transfers, e.g., a few KBytes. The reason is that MIMO requires collecting detailed channel state information which introduces overhead. A second factor is that not all wireless devices have the same level of MIMO support. For example, WiFi basestations tend to have a more radios and antennas than client devices. The reason is that client devices have more severe size, cost, and power constraints than basestations. Unfortunately, the number of parallel streams is limited by the device with the lowest number of antennas, so increasing the number of radios and antennas on access points will in practice have limited benefit.

Multi-User MIMO (MU-MIMO) is a variant of MIMO that is designed to allow access points to fully benefit from MIMO, even when communicating with resource-constrained devices (Spencer et al., 2004). We will use Fig. 3.13 to illustrate the concept. Figure 3.13a shows an example of an AP and one client device, each with four antennas, communicating using MIMO with four data streams, as discussed in Sect. 3.9. Figure 3.13b shows a similar MU-MIMO example of an AP with four antennas communicating with two clients, each with two antennas, simultaneously receiving two data streams from the AP. Not surprisingly,

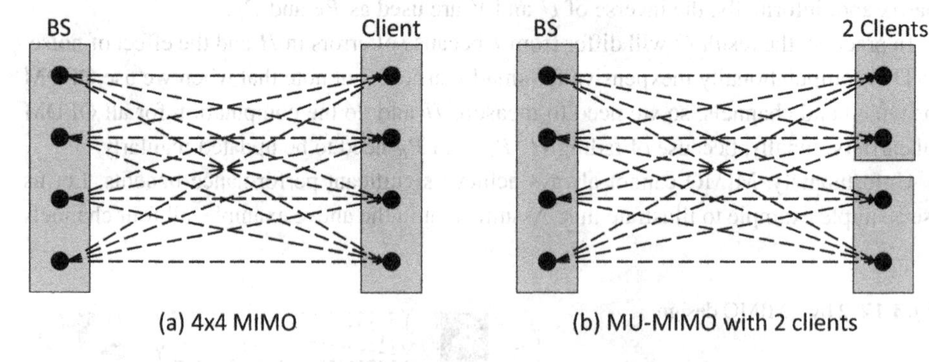


**Fig. 3.13** Comparison of MIMO and MU-MIMO

MU-MIMO is more challenging to implement than MIMO. In both cases, each receiving antenna receives a signal that is a combination of the four transmitted signals. In the case of MIMO, the receiver has four signals that it can use to extract the data streams, but with MU-MIMO, it has access to only two signals, so it has less information that it can use to extract the two signals that correspond to its two data streams. As a result, the aggregate throughput with MU-MIMO will often be lower than with MIMO, but MU-MIMO has the advantage that multiple clients can be served at the same time.

MU-MIMO can be used for both uplink and downlink communication. For downlink communication, the received signals will be properly synchronized, similar to MIMO. Uplink communication is more challenging, since the data streams are transmitted by different devices. In order for uplink MU-MIMO to work well, the transmissions by the two transmitters must be synchronized very precisely.

## 3.10 Spread Spectrum

In Sect. 3.6, we describe how OFDM uses time and frequency diversity to address the challenges associated with frequency-selective fading. However, OFDM is a fairly expensive technology so it is only appropriate for higher speed links. In this section we introduce spread spectrum technology as an alternative technology to deal with frequency-selective fading.

### 3.10.1 Spread Spectrum Concept

The easiest way to explain spread spectrum (SS) is to consider Shannon's Law (Eq. 2.2), which we repeat here for convenience:

$$C = B \times log_2\left(1 + \frac{S}{N}\right) \tag{3.12}$$

$C$ is the maximum capacity(bit rate) of a channel of width $B$ and a signal-to-noise ratio $S/R$ at the receiver. So far, we have focused on technologies that optimize $C$, given a certain bandwidth $B$. However, many devices have low bandwidth requirements, but they must be low cost and power efficient. Shannon's law shows that if we want to achieve a certain bit rate $C$, then the simplest way to do this is to use a lot more bandwidth $B$ then we strictly need, so the communication will be very robust. The benefit is that we can use simple radios and still meet our target bit rate, even if noise levels are high, or if the signal is distorted as a result of multi-path effects and mobility. There are two classes of spread spectrum technologies: Frequency Hopping (FPSS) and Direct-Sequency (DSSS) Spread Spectrum.

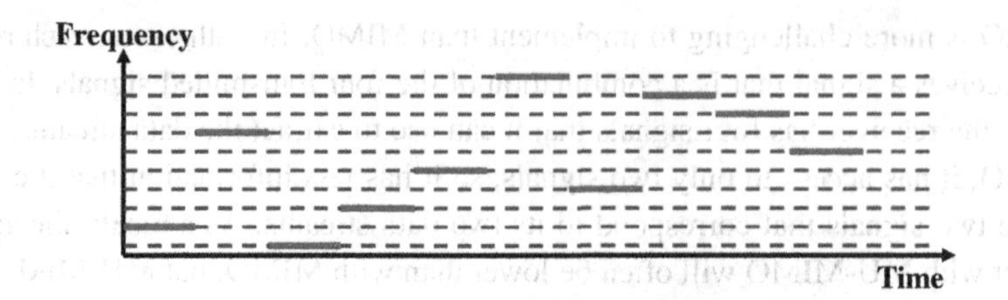


**Fig. 3.14** Frequency hopping spread spectrum

### 3.10.2 Frequency Hopping Spread Spectrum (FHSS)

Figure 3.14 shows an example of FHSS. The frequency band that is available for transmission is partitioned into several narrow subchannels (y-axis). Each of the channels can be used to transmit a narrowband signal with a low bit rate. The transmitter transmits on one channel at a time, hopping around between channels over time (x-axis). Frequency hopping SS is very robust in multi-path environments since it supports time and frequency diversity. The reason is that in any time slot, only a small number of the channels will suffer from strong attenuation and what channels are affected changes over time. This means that in a given time window the transmissions in most slots will be successful, and forward-error correction can be used to recover the data sent on subchannels with high attenuation. The sequence of channels that is used for transmission is called the *hopping sequence*. Hopping sequences are defined by the standard and the sender and receiver must use the same hopping sequence so they remain in sync.

FHSS is used by Bluetooth in the 2.4 GHz ISM band. It uses 79 channels with a spacing of 1 MHz. The hopping rate is 1600 hops/sec. The original standard has a bit rate of 1 Mbps, which was increased to 3 Mbps in version 2.1 There is also a low power version of Bluetooth called Bluetooth Low Energy (BLE) which is used in IoT devices.

### 3.10.3 Direct Sequence Spread Spectrum (DSSS)

Figure 3.15 shows the steps involved in communication using DSSS. The first row shows the user data stream that must be transmitted. Each bit is expanded into 3 bits, sometimes called a *chip*, by doing an exclusive OR with a 3 bit *spreading code*, adding redundancy to the data stream. The resulting bit stream is then used to modulate the carrier (we show amplitude modulation for simplicity). Modulation uses the full bandwidth B of the spectrum band. The bottom half of Fig. 3.15 shows that the receiver executes the same steps but in the reverse order: the signal is demodulated, and each chip is converted into a bit using an exclusive OR with the spreading code.

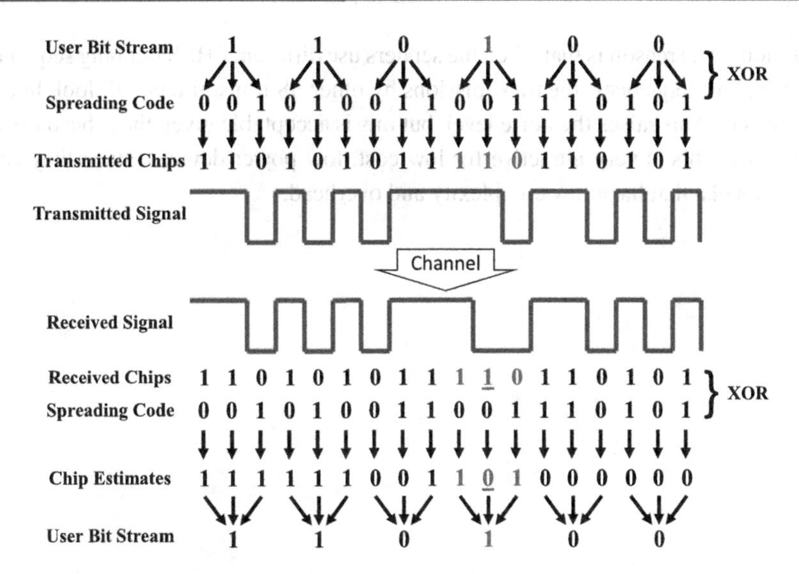


**Fig. 3.15** Direct-sequence spread spectrum

To illustrate how error correction works, we have introduced a bit error in the middle of the bit stream, resulting in a corrupted chip (highlighted in red in the figure). For chips that were received correctly, the XOR operation will result in three bits that are identical to the bit value that was used to generate the chip. When there is a bit error, one of the bits will have a value that is different from the original bit value. Simple voting is then used to determine the actual bit value. In this simple example, DSSS can recover from a single bit error per transmitted chip. By using more spectrum than is need, in addition to redundancy, communication based on DSSS is very robust, even in the presence of frequency selective fading and noise. Clearly robustness can be improved by using more chips per bit, at the cost of using more spectrum. DSSS was used by some 3G cellular technologies. Today, it is widely used in IoT protocols. One example is Lora, a physical layer used by several IoT protocols, including LoraWan (Li & Cao, 2022).

### 3.10.4 Multiple Access Based on Spread Spectrum

So far, we have described spread spectrum as a physical layer technology that improves the robustness of wireless transmission. However, spread spectrum can also be used as a multi-access technology that allows multiple transmitters to share the same frequency band. As we discuss in the next chapter, the sharing of a spectrum band by multiple transmitters generally requires protocol support to coordinate the transmissions by the transmitters. Spread spectrum technology offers an alternative. Since spread spectrum transmissions are very robust, multiple transmitters can share a frequency bandwidth without explicitly coordinating their transmissions. The reason is that when the senders use different FHSS hopping sequences or DSSS spreading sequences, the transmissions by other SS transmitters will look like noise to SS receivers. This raises the noise level, but this is acceptable given the robustness of the transmissions. This is very attractive for low cost, low power devices, since they can use simple protocols, that have low complexity and overhead.

# 4 Wireless Datalink Protocols

## 4.1 Why so Many Wireless Protocols?

The two protocol families we will discuss represent only a tiny fraction of the wireless protocols in use today. This raises the question why we need so many wireless protocols. There are many reasons:

- Wireless devices often have different goals and constraints. Examples are throughput, power consumption, cost, physical environments (indoor/outdoor, range, ...), etc.
- As described in Sect. 2.1, datalink protocols operating in the unlicensed spectrum need to be able to co-exist with other networks that may use either the same protocol or a "foreign" protocol. This typically requires protocol features that introduce extra overhead. In contrast, datalink protocols in the licensed spectrum can be more efficient since they do not have this constraint.
- Over time the PHY layer technologies have improved dramatically, often requiring changes to the datalink protocol to support them. Examples are the need to collect fine grain CSI or the use of packet bursting to amortize overheads. Similarly, technology improvements have made it possible to use higher frequency bands in the spectrum, such as mmWave, which can also require changes to datalink protocol.

Of course, wired datalink protocols also have had to change over time, mostly to support newer PHY technologies that increase throughput. However, these changes have been very limited, in part because the signals propagate in an isolated transmission medium (the wire) and the protocols are much simpler.

In this chapter we describe protocols in two wireless protocols families. Specifically, we discuss IEEE 802.11 (WiFi) and cellular, operating in the unlicensed and licensed spectrum respectively. These two technologies have historically quickly introduced the newest wireless

 
P. Steenkiste, *Introduction to Wireless Networking and Its Impact on Applications*,
Synthesis Lectures on Mobile & Pervasive Computing,
https://doi.org/10.1007/978-3-031-27466-4_4

physical layer optimizations to maximize throughput. As a result, these technologies tend to be relatively expensive and power hungry. We will describe these wireless datalink protocols at a fairly high level. Our focus is on features that differ from wired protocols or that differ across different wireless protocols. The goal is to help users decide when the various technologies are appropriate and identify factors that may impact performance.

## 4.2 Datalink Responsibilities

The datalink layer is responsible for the transfer of packets between nodes that are connected to the same layer 2 network. A simple example is a homogeneous network based on switched ethernet, as is widely used in data centers, campus networks, etc. A more complex example is a network that has links based on multiple 802.11 standards (e.g., Ethernet and WiFi), that are designed to interoperate at layer 2, i.e., without needing help from layer 3 (the Internet protocol). Note that datalink Protocol Data Units (PDUs) are technically called "frames", while the term "packet" is used for network (IP) PDUs, although both are often informally referred to as packets.

As described in Chap. 2, the datalink layer receives requests for packet transmission from the network layer, specifically a sequence of bytes that have to be sent to a specific destination (Chap. 1). To deliver the data, the datalink layer can rely on the help of the physical layer to send a sequence of bits between two nodes. The general responsibilities of the datalink layer falls in three areas: framing, logical link control, and media access control.

### 4.2.1 Framing

Framing is responsible for encapsulating the data to be transmitted into a frame, which includes a header and a trailer. The header contains the addresses of communicating devices and control information, such as packet type (e.g., data or one of several control types) and service type (e.g., unicast or broadcast). For wireless networks, it may also include information needed by the receiver to demodulate, decode, and interpret the frame (see Sect. 4.3.7 for an example). The trailer at the end of the packet typically contains a checksum, as described below. Framing is an important function but similar techniques are used in wired and wireless networks, so we will not further elaborate on this function.

### 4.2.2 Logical Link Control

*Logical link control* is a set of functions that require coordination between the sender and receiver. The two main functions are (1) flow control and (2) error detection and correction.

The flow control function prevents the sender from sending frames faster than the rate at which the receiver can process and store them. It is implemented by having the receiver tell the sender regularly how many frames it is allowed to send, e.g., based on available buffer space. Flow control is mostly supported for network technologies that involve high speed links, and it is not common in wireless.

### 4.2.3 Error Detection and Recovery

Error detection and correction, sometimes referred to as error control, is a very important function in any datalink protocols and since wireless links tend to have higher error rates than wired links, they tend to have more sophisticated error detection and correction protocols. A first requirement for error control is that the receiver can determine whether a received frame is correct, i.e., identical to the one sent by the sender. This is achieved by appending an Error Detection Code (EDC) to the data frame as a trailer.

$$S_{wire} = S_{data} + S_{EDC} \tag{4.1}$$

Where $S_{wire}$ is the size of the transmitted packets, $S_{data}$ is the size of the data (user payload plus protocol headers), while $S_{EDC}$ is the size of the error detection code. The EDC is calculated by the sender using the original data frame and it allows the receiver to check, with high probability, the correctness of the transmission by calculating the EDC on the received data frame and comparing it with the EDC in the trailer. A typical EDC used in packet networks is 32 bits long, so the overhead is very low. Packets with errors are discarded. Datalink protocols generally use a Cyclic Redundancy Check (CRC) as to detect errors. It is based on polynomial division and it is cheap to implement in hardware at line rates.

It is important to understand that no EDC can guarantee with certainty that a transmission is correct. The intuition is that a data frame plus EDC, sometimes referred to as a code word, can be valid or invalid, depending on whether the CRC matches the data. When a sender transmits a code word, it is always possible that bit errors resulting from the transmission will result in a received code word that is different but also valid. One metric for evaluating EDCs is its *Hamming distance* (Hamming, 1950), which is the smallest number of bits that, when changed, maps one valid code word into another valid code word. The Hamming distance depends on the type of EDC, the length of the data frame and EDC, and the type of errors (e.g., random bit errors versus error bursts). An important advantage of CRC codes is that they have been studied extensively, so their error detection properties are well understood (Koopman, 2002).

### 4.2.4 Error Recovery

Whether datalink protocols recover from corrupted or lost packets depends on the protocol. In this section we first describe error recovery at the datalink layer and then we discuss the

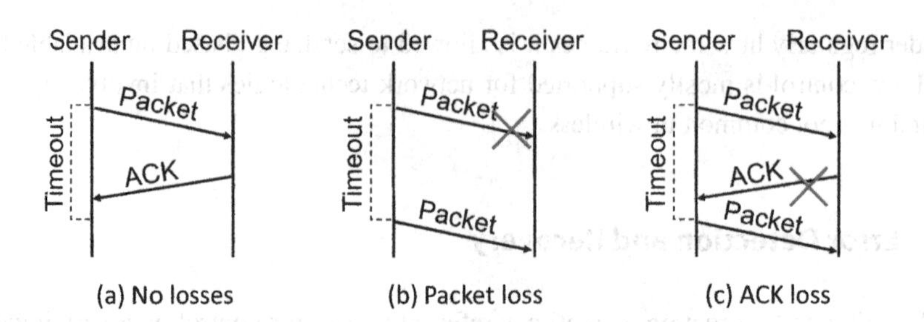


**Fig. 4.1** Error recovery using retransmission

tradeoffs between error recovery at the datalink layer and end-to-end recovery by a transport protocol such as TCP.

Two classes of techniques are used to recover from lost packets. The first solution is the *retransmission of the frame*. While many variants exist, datalink layers tend to use a simple solution based on acknowledgements (ACKs). Figure 4.1 shows a simple example. After every successful packet reception, the receiver sends an ACK to the sender. If the sender does not receive an ACK within a certain timeout after the transmission, it retransmits the packet. Since wireless links tend to be short, the timeout interval tends to be short, so error recovery is fast. In order to deal with different error scenarios, such as lost ACK frames, packets must have sequence numbers so the receiver can verify that it received all packets, in the correct order, without duplicates.

Retransmission is easy to implement, and it is very effective when the frame loss rate is low. However, it can be very inefficient, for example, when frames have just a few incorrect bits. *Forward Error Correction* is an efficient way to recover packets that are mostly correct. The idea is similar to the use of error detection codes. The original frame is extended with extra bits based on an Error Correction Code (ECC). However, while EDCs can only detect errors, ECCs can, in addition to detecting errors in corrupted code words, also correct the error for a subset of the corrupted code words. Similar to EDCs, ECCs cannot guarantee that code words received with a correct ECC, or corrected after the ECC failed, are always correct (Moon, 2005). Not surprisingly, the use of ECCs generally adds more overhead than EDCs.

Not all datalink protocols implement error recovery. Wired networks such as Ethernet have extremely low error rates and lost packets are not retransmitted. Instead, applications that need reliable communication need to use end-to-end recovery using a transport protocol such as TCP or possibly at the application level. Since error rates are low, the per-packet overhead associated with error recovery (ACKs or error-correcting codes) cannot be justified. It is more efficient to rely on end-to-end error recovery using a transport protocol such as TCP for the occasional packet loss. Wireless datalink protocols typically implement local error recovery. This solution is preferred since error recovery at the datalink layer is much more efficient than end-to-end error recovery. First, datalink level recovery only increases overhead (ACKs or the use of ECCs) on one link rather than on every link on the end-to-end path. Second, when retransmission is used, recovery is much faster since roundtrip times are much shorter at the link level than end-to-end. With high link-level error rates, the cost of end-to-end recovery far exceeds that of local recovery.

This approach to error recovery is consistent with the end-to-end principles (Saltzer et al., 1984), which states that functions such as error recovery should be implemented end-to-end, unless there is a clear benefit to providing support inside the network. Local recovery on links with high error rates are a good example. Note that for applications that require reliable communication, end-to-end error recovery must still be used, even when local error recovery is used for lossy links in the Internet. The reason is that end-to-end recovery covers *all* the links, in addition to the routers and switches on the network path.

### 4.2.5 Media Access Control

Media access control is the function in the network that determines which node can use the transmission medium to transmit a packet. The need for this function depends on the type of datalink architecture that is used.

Figure 4.2 shows two common architectures. Today's wired datalink technologies typically use a *store-and-forward architecture* (Fig. 4.2a). It consists of a set of switches (squares) connected by links providing network connectivity for a set of client nodes (circles). The packets transmitted by client devices are forwarded hop-by-hop by the switches, based on the layer 2 address, towards the destination node. A common example today is switched Ethernet. Today, links based on wires are typically full duplex, which means that both devices connected to a link can simultaneously transmit and receive over the link. With full duplex links, devices can send at any time and no media access control is needed.

In the second architecture, called *multiple access networks*, multiple devices share the same transmission medium (Fig. 4.2b). The assumption in multiple access networks is that packets transmitted by any device are received by all potential receivers in the network. Individual receivers either process the packet and pass it up the protocol stack, or discard it, depending on whether they are the intended recipient. Examples include older local area

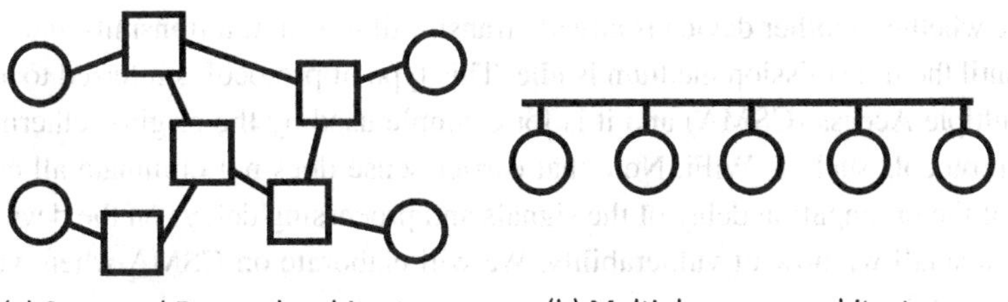

(a) Store and Forward architecture
(b) Multiple access architecture


**Fig. 4.2** Datalink network architectures

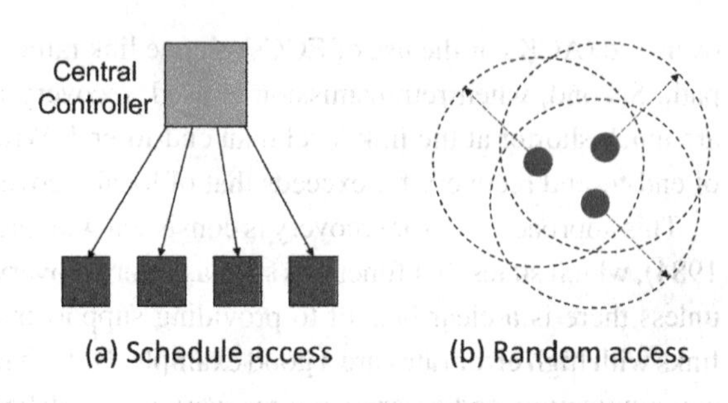


**Fig. 4.3** Scheduled and random access MAC protocols

network technologies, in addition to wireless. Since two or more potential transmitters share the same transmission medium, some form of coordination is needed to avoid simultaneous transmission. In general, simultaneous transmissions means that one or more potential receivers will receive overlapping signals, so it may not be able to demodulate and decode the packet. This is called a *collision*.

Multiple access networks fall in two categories, depending on how they avoid, or minimize, collisions. In *scheduled access networks*, nodes explicitly coordinate. Figure 4.3a shows an example in which a centralized controller controls channel access. Nodes can only transmit after they have received permission from the controller, as illustrated by the black arrows. Scheduled access requires a control channel supporting communication between the clients and the controller. Alternative designs use a token that is passed between the nodes; a node can only transmit when it has the token. Schedule access does not only have the advantage that it eliminates collisions, but it also provides control over how bandwidth is allocated to clients or classes of traffic (e.g., video streaming versus web browsing). The disadvantage is that coordination introduces overhead, since the control traffic takes away bandwidth from data traffic. We will discuss an example of scheduled access in more detail when we describe cellular technologies in Sect. 4.4.

In *random access networks*, there is no explicit coordination in the form of control messages between the nodes. Figure 4.3b shows an example of three wireless nodes that are within communication range of each other, as illustrated by the dashed circles. Clearly if the nodes transmit packets at random, there will be a lot of collisions when the load increases. A simple technique to avoid this is to "listen before you talk", which is similar to how people coordinate in a meeting. Before transmitting a packet, the device listens to determine whether another device is already transmitting. If not, it transmits and otherwise it waits until the transmission medium is idle. This type of protocol is referred to as Carrier Sense Multiple Access (CSMA) and it is for example used by the original ethernet and in wireless protocols such as WiFi. Note that carrier sense does not eliminate all collisions. Because of the propagation delay of the signals and processing delays on the devices, there is always a small window of vulnerability. We will elaborate on CSMA when we discuss WiFi in Sect. 4.3.

## 4.3 WiFi: High Performance in Unlicensed Spectrum

In this section we introduce the high-level architecture of WiFi. We then review the features of the major WiFi versions and present some of the primary control plane functions. Finally, we discuss some key WiFi features relevant to users and a few recent WiFi versions that were developed for specialize deployments. We first provide some background on WiFi's history and standardization process.

#### History

WiFi was enabled by a 1985 ruling by the US FCC, which designated the ISM bands for unlicensed use (Sect. 2.1.2). The early ideas for WiFi were developed by NCR corporation (later acquired by AT&T) in 1991 as a way of connecting cashier systems wirelessly to a local area network. Soon after that, the IEEE established the 802.11 working group for WiFi LAN technologies, with the goal of giving wireless devices the same type of access to the Internet as wired devices. During the nineties, the 802.11 working group developed the first two versions of WiFi, 802.11 b and a, with a maximum bit rates of 11 and 54 Mbps respectively. Since then, new versions of WiFi have been released regularly, incorporating the latest physical layer technologies, in addition to a number of protocol improvements. The early versions of WiFi used the 2.4 and 5 GHz unlicensed bands, while newer versions can also operate in the 60 and 6 GHz bands, which were made available by the FCC in 2014 and 2020 respectively.

#### A Word About Standards

The IEEE is responsible for many standards, include most standards for local area networks in use today. LAN standards fall in the 802 group. The standards for each protocol family are defined by a working group which is define by a number. For example, Ethernet is 802.11.3, WiFi is 802.11, and personal area networks (PAN) are 802.15. Each protocol family consists of many standards, each defining protocol versions and various extensions. These standards (and the working groups defining them) are identify using the protocol family number followed by a letter. Letters are assigned in the order in which the working groups were established. After 26 working groups, two letters are used. For example, the two latest WiFi versions are 802.11 ax and ay.

### 4.3.1 High Level WiFi Architecture

WiFi network technology can be used in two operating modes that differ in how nodes use the WiFi standard to communicate.

First, *ad hoc mode* is used to build store and forward WiFi networks, not unlike the store and forward architectures illustrated in Fig. 4.2a. There are however substantial difference

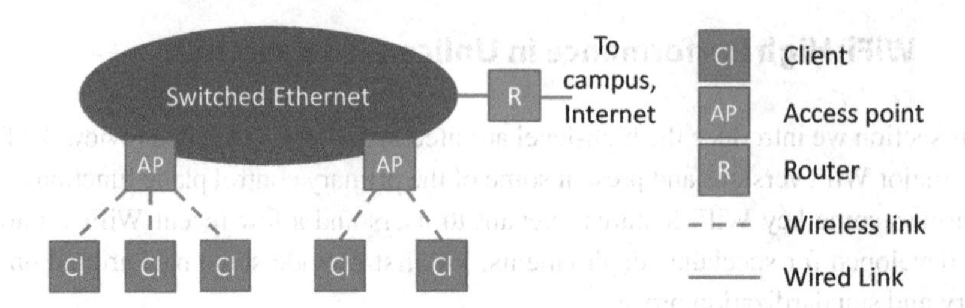


**Fig. 4.4** WiFi infrastructure mode

with, for example, switched Ethernet. First, in switched Ethernet, switches are customized devices that are maintained by a professional staff, while in ad hoc networks, all WiFi nodes tends to be equals, acting as both clients and switches. Second, in contrast to switched Ethernet, the transmissions on the "links" connecting the ad hoc nodes can interfere with each other, resulting in reduced bandwidth and high packet loss rates, as we discuss in more detail in Sect. 4.3.2. Finally, in many applications considered for ad hoc networking, such as Vehicular Ad Hoc Networking (VANET), the ad hoc nodes are mobile, resulting in very dynamic topologies. Because of these challenges, there is a large body of research on WiFi "ad hoc" networks. However, in practice, they are rarely used and WiFi ad hoc mode is not supported in more recent WiFi versions. While multi-hop wireless networks are used in some specialized deployments (mesh networks, some military and first responder networks), we will not consider them further.

The second mode is called *infrastructure mode*. Infrastructure mode is widely used to deploy WiFi LANs, e.g., in homes, hotspots, campuses, etc. It consists of a set of basestations, each providing a set of wireless devices with access to the wired network infrastructure, as shown in Fig. 4.4. This simple network shows a LAN with two basestations (BS) serving client devices (Cl). The two BSs are connect to a switched ethernet, which may also serve wired clients as part of a larger LAN infrastructure. Finally, the LAN is connected using a router to either a larger campus network infrastructure, or directly to the Internet. Home networks often consist of a single basestation that is directory connected to a last mile ISP.

Each basestation and its clients form a single WiFi network that is called a Basic Service Set (BSS). All wireless communication happens between the basestation and a client, i.e., clients do not communicate directly with each other. This may seem strange, but remember that the focus of WiFi is to connect wireless devices to the Internet. Today, there are some technologies such as WiFi Direct that allow WiFi devices to communicate directly without a basestation, but these direct links are not part of the LAN. The use of basestations simplifies network management since basestations are controlled by network managers.

In a campus networks, a group of BSSes that are connected to the same LAN are managed as a single network, which is referred to as an Extended Service Set (ESS). Since WiFi and switched Ethernet are both IEEE standards, they interoperate. Specifically, the basestation effectively acts as a layer 2 switch, so clients can use the basestation to communicate at layer 2 with other WiFI and ethernet clients in the same BSS or ESS. Basic and Extended Service Sets have an Service Set Identifier (SSID), which is a name that allows devices and their users to identify basestations they want to connect to, e.g., Bobs-WiFi or CMU-secure (see Sect. 4.3.9).

Multiple access networks, such as a WiFi BSS, can either use either schedule access or random access MAC (Fig. 4.3). Considering that in a BSS all communication is between a basestation and its clients, scheduled access, as shown in Fig. 4.3a, seems like a natural fit and there is in fact a standard for a scheduled access MAC in WiFi, called Point Control Function (PCF). However, as discussed in Sect. 4.2.5 scheduled access incurs a significant overhead, so while PCF is part of the standard, it is rarely used. WiFi deployment use random access, called the Distributed Control Function (DCF), as illustrated in Fig. 4.3b. Interesting enough, this means that while the basestation clearly has a specialized role in a BSS, from the perspective of media access control, all WiFi devices, including basestations and clients, are equals.

### 4.3.2 Wireless MAC Challenges

In wired networks such as switched Ethernet, devices including both switches and clients are connected by point-to-point wires that capture the EM signal as it travels between the transmitting and receiving radio. This results in low attenuation and it also isolates the signal from other EM signals in the environment (noise, interference). In contrast, in wireless networks, signals travel freely in a shared physical environment. This difference impacts wireless signal propagation and communication in three ways. First, wireless channels are often shared by many transmitters that must coordinate when communicating. Second, wireless channels have higher attenuation and noise than copper wires or fiber. Finally, unless coordination between senders is perfect, transmissions can interfere with each other. This section introduces the main WiFi features that address these challenges.

### 4.3.3 Wifi Protocol Design

In this section we describe how the WiFi protocol addresses the above wireless challenges. Another protocol design constraint is that WiFi is used in unlicensed spectrum band, so it needs to coexist with other protocols.

Simultaneous transmissions by nearby devices using WiFi or other protocols will "collide", meaning that the intended receiver receives overlapping signals that can typically results in packet loss. To avoid collisions, WiFi uses *carrier sense*. As described in Sect. 4.2.5, carrier sense is easy to implement and can avoid many collisions with minimal overhead. Unfortunately, carrier sense cannot eliminate all collisions. Senders can start a transmission at roughly the same time (e.g., within 10 s or 100 s of nsec), which means that a second sender cannot detect the other transmitter before they send given processing and signal propagation delays.

The high attenuation on wireless channels can also lead to carrier sense failures and collisions. As discussed in Sect. 2.5, attenuation is proportional to $d^n$ at distance $d$ (Eq. 2.3), or $n \times log(d)$ in dB, given a path loss exponent $n$. In contrast, on wires, attenuation is exponential in the length of the wire. Specifically, the attenuation in dB is $k \times d$ where $k$ is the loss per meter (unit of $d$). This would suggest that wires have higher attenuation than wireless channels, but the constants matter. Attenuation in wireless, for example for WiFi at 2.4 GHz, can be in the range of 60–100dB because the antenna radiates energy in many directions. In contrast, attenuation on a wire of a similar length can be on the order of 1 dB or less, depending on the type of wire that is used, so it does not impact carrier sense.

High attenuation in wireless can lead to carrier sense failure because wireless devices may only be able to hear (and talk to) to a subset of nodes, while other nodes may be out of range. As a result, when a sender performs carrier sense, it may fail to detect that the intended receiver is already receiving a transmission from another sender. One way to avoid this problem is to limit the distance between all nodes sharing a spectrum band to a conservative, maximum distance, thus limiting attenuation. Unfortunately, this is not practical in a wireless network, especially with mobile nodes. Carrier sense failures due to attenuation can lead to two problems.

A first problem is called *hidden terminals*. Figure 4.5a shows a simple example with three nodes. We see that nodes A-B and B-C are within range of each other, but A-C are not. When A sends a packet to B, carrier sense will not prevent C from transmitting a packet to B at the same time, resulting in a collision. This scenario can for example happen when two clients A and C are on opposite sides of a basestation. Figure 4.5b shows a more general case. When A is transmitting to B, any transmission by C, independent of the intended destination, will result in a collision at node B. This can for example happen when B and D are basestations using the same frequency band in an infrastructure deployment.

The second problem is called *exposed terminals*. Figure 4.6 shows an example with four nodes. When B is transmitting a packet to A, carrier sense will prevent node C from simultaneously sending a packet to node D. Note however that D cannot hear B's transmission, so

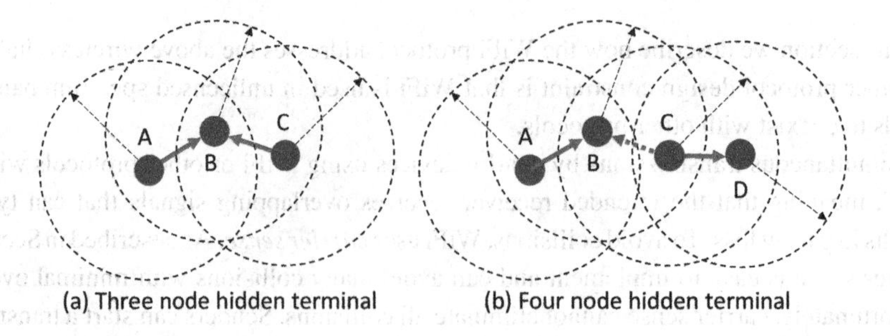


**Fig. 4.5** Hidden terminal scenarios

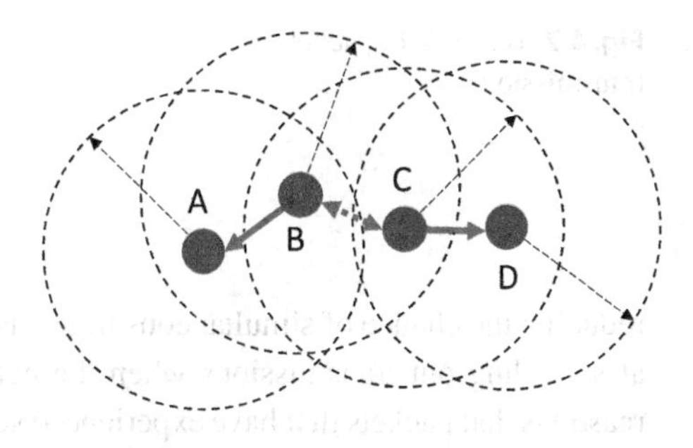


**Fig. 4.6** Exposed terminal scenario

it would be able to receive packets from C without a problem, but carrier sense needlessly prevents the transmission. In contrast to hidden terminals which result in collisions, exposed terminals reduce network capacity by preventing transmissions that are likely to succeed.

The above examples assume free space communication. Adding obstacles increases the likelihood of hidden and exposed terminals since even nodes that are physically close may not be able to hear each other. In addition, due to fading, which nodes can hear each other changes over time, possibly even during packet transmission. These factors impact the effectiveness of carrier sense. However, overall it is still a big win since it has very low overhead and avoids most collisions, so carrier sense is an important feature of the WiFi protocol. Also, as we discuss in more detail in Sect. 4.3.10, hidden and exposed terminals are generally not a big problem in infrastructure WiFi deployments.

Given that collisions cannot always be avoided, it is important to minimize the performance impact of collisions. Ideally, when a collision happens, the two senders will be able to detect the collision right away, so they can (1) abort their transmissions to avoid wasting time sending useless signals and (2) retransmit the packet. In wired random access network, a feature called *collision detection* allows transmitters to detect collisions immediately so it they can recover quickly. Unfortunately, collision detection is not possible in wireless because of the high attenuation of the wireless link. This means that the cost of collisions in wireless is high, since transmitters of the colliding packets will complete their transmissions, not knowing that their packets will be lost. This is one of the reasons why wireless protocols such as WiFi implement error recovery, so lost packets are automatically retransmitted after a timeout.

After a collision, the two nodes need to retransmit their packet. However, if both nodes retransmit their packet right away, there is likely to be another collision. *Random backoff* tries to avoid this by having the nodes wait a random amount of time before retransmitting after a packet loss. Random backoff is implemented by having nodes wait a random number of time slots, where the number is picked from a backoff window. For example, if the window is 4, each node generates a number in the range 0–3 and waits for that many slots. The node that picks the lowest number will be able to transmit. The other node will in most cases detect the transmission and wait. If there is another collision, e.g., because the nodes picked the same number or there is a hidden terminal, the nodes double the back off window, further

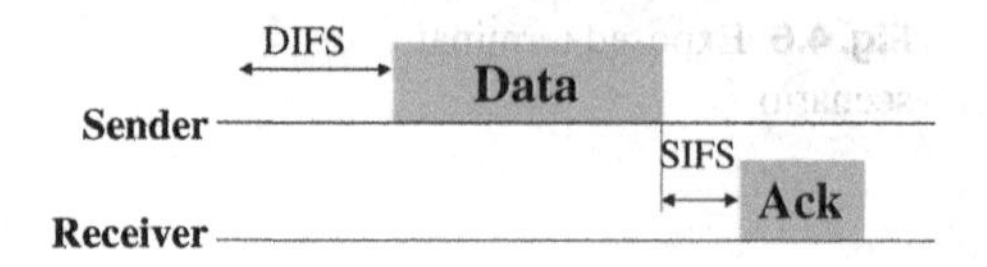


**Fig. 4.7** Basic WiFi packet transmission

reducing the chance of simultaneous transmissions. Such "exponential" backoff is effective at spreading out transmissions when the network is congested, but it can be unfair. The reason is that packets that have experienced several collisions have a longer backoff window than new packets, reducing their chances of gaining access to the channel when other nodes have shorter backoff windows. This can lead to long and highly variable packet delays.

Finally, WiFi uses variable bit rate so it can optimize throughput based on the channel conditions, as discussed in Sect. 3.1. Changes in bit rate directly impact the network performance of applications.

### 4.3.4 The Basic WiFi Protocol

The basic WiFi protocol is very simple, as shown in Fig. 4.7. A sender can transmit a packet after it has used carrier sense to check that the channel is idle. If the packet is successfully received, the receiver sends an acknowledgement. If the sender does not receive an ACK, it retransmits the packet after a random back off, as discussed in the previous section.

All packets transmitted in WiFi are separated by an *inter-frame spacing* (IF). In the figure we see that different interframe spacings are used. Before the data packet is sent, the sender has to wait for a DCF Interframe Spacing (DIF), while before the ACK is sent, the receiver waits for a shorter Short Interframe Spacing (SIF). The reason for the shorter spacing is that it guarantees that ACKs have priority over data packets, so they are always sent right after the packet they are acknowledging. This simplifies the protocol considerable.

In order to improve the performance of carrier sense, WiFi also implements a second protocol-level carrier sense mechanism called *Virtual Carrier Sense* (VCS). The "listen before you talk" carrier sense is referred to as physical carrier sense (PCS). The idea is to embed information in the packet header that specifies how long the transmission of the packet, including the ACK, will take. The sender can calculate this information based on the length of the packet, the bit rate, and the duration of the SIF and ACK. It stores this information in the packet header so any node that hears the header knows how long it has to defer, even if it only hears the header, for example, as a result of fading. Because of the importance of the VCS and other information in the header, the header is transmitted at a lower bit rate than the data. This not only increases the chance that nearby nodes will receive the VCS information, but it also may allow more remote nodes to receive it. This is illustrated in Fig. 4.8. WiFi devices implement VCS using a Network Allocation Vector (NAV), which has an entry for each device that has an active transmission listing when its transmission ends.

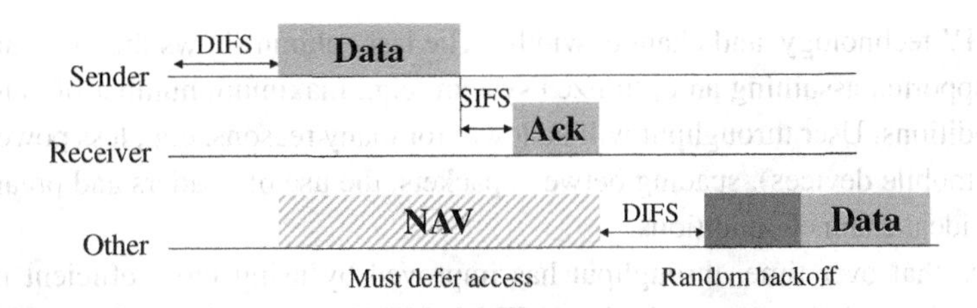


**Fig. 4.8** WiFi virtual carrier sense

Note that virtual and physical carrier sense have slightly different properties. PCS will defer to any signal, including transmissions by other protocols (and even microwaves), while VCS only defers to other WiFi devices. The default setting for WiFi is often to only use VCS.

The WiFi standard also defines a *collision avoidance* (CA) feature which is designed to prevent collisions due to hidden terminals. With CA, the sender sends a short Request to Send (RTS) packet to the receiver, asking for permission to send, to which the receiver responds with a Clear to Send (CTS). Both packets include the VCS information for the transmission (including control packets), so all nodes near both the sender and the receiver will defer to the transmission. CA incurs significant overhead since it requires four packet transmissions per data packet, so it is not used in practice.

### 4.3.5 WiFi Versions

Table 4.1 summarizes key features of the various WiFi versions. The columns compare the following features for each standard: release date, spectrum bands used, the primary features

**Table 4.1** Main features of WiFi version

| Version | Date | Bands | PHY features | Channels (MHz) | Max (Mbps) |
|---|---|---|---|---|---|
| 802.11b | 1999 | 2.4 | DSSS | 20 | 11 |
| 802.11a | 1999 | 5 | OFDM | 20 | 54 |
| 802.11g | 2003 | 2.4 | OFDM | 20 | 54 |
| 802.11n | 2009 | 2.4, 5 | OFDM, MIMOd, BF | 20, 40 | 450 |
| 802.11ac | 2013 | 5 | OFDM, MU-MIMOd, BF | 20,40, 80, 160 | 3.5 Gbps |
| 802.11ad | 2016 | 60 | BF | 2.16 GHz | 6.8 Gbps |
| 802.11ax | 2021 | 2.4, 5, 6 | MU-MIMO, BF, OFDMA | 20, 40, 80, 160 | 9.6 Gbps |
| 802.11ay | 2021? | 60 | MU-MIMO | 2.16, 4.3, 8.6 GHz | 20 Gbps |

of the PHY technology, and channel widths. The last column shows the maximum bit rate that is supported assuming an optimized system (e.g., maximum number of antennas) and ideal conditions. User throughput will be lower for many reasons, e.g., less powerful radios (e.g., on mobile devices), spacing between packets, the use of headers and preambles, and less than ideal channel conditions.

We see that over time, throughput has improved by using more efficient modulation and other physical layer technologies (OFDM, MIMO). In addition, the maximum number of antennas supported has increased (not shown). Features such as MU-MIMO and OFDMA also provide more fine grain distribution of bandwidth across clients, as described in Sects. 3.9.2 and 4.4.5. Note that these PHY technologies require changes to the WiFi protocol since they support communication between a basestation and a group of clients while the original WIFi protocol is designed for point-to-point communication (Sect. 4.3.4). Finally, more recent versions of WiFi support wider channels, which according to Shannon can increase throughput linearly. The cost is that this decreases the number of non-overlapping channels that can be supported in a frequency band. For example, the 2.4 GHz band supports three 20 MHz non-overlapping channels, but only one 40 MHz channel and no wider channels. In contrast, the 5 GHz band has well over 200 MHz of bandwidth, although details of availability and usage constraints differ across countries.

The increased throughput in the more recent WiFi versions comes of course at a cost. OFDM is computationally expensive, with the cost increasing with the number of subcarriers. MIMO requires one radio per antenna and has a processing cost on both the sender and receiver that is quadratic in the number antennas. Wider bands also use more OFDM subcarriers, further increasing cost. Finally, as technologies become more sophisticated, they need more fine grain CSI. This is obtained by sending preambles or control packets, which consume transmission time, further increasing overhead.

### 4.3.6 mmWave WiFi

802.11ad and ay are the first WIFi protocols operating in the 60 GHz band (mmWave). The mmWave band is very challenging for wireless communication. First, the free space path loss equation (Eq. 2.3) shows that attenuation increases as $f^2$ where $f$ is the frequency. This means that attenuation in the 60 GHz band is about 140 times higher than in the 5 GHz band and 625 higher compared to the 2.4 GHz band. In addition, mmWave signals are blocked by obstacles. Finally, absorption of RF signal energy in the mmWave bands is higher due to water vapor and oxygen in the atmosphere. This results in very short communication ranges. Early, pre-standard mmWave networks, for example, had a range of 10 m or less, so their use was limited to room size systems with point-to-point high bandwidth requirements.

There is however also good news. First, there is a lot of mmWave spectrum. The FCC has for example allocated 14 GHz of unlicensed spectrum for broadband used. Most mmWave WiFi devices support four non-overlapping channels, each with 2.16 GHz of spectrum.

Second, mmWave antennas are very small due to the high frequency, so it is possible to use a lot of antennas, even on small devices such as cell phones. This means that beamforming using antenna arrays is an attractive option. Since the width of the beam is roughly inverse proportional to the number of antennas, very narrow beams are possible, extending the communication range. For example, 80211ad can have up to 32 antennas.

The aggressive use of beamforming in the mmWave band comes however with significant challenges. A first challenge is how clients and a basestation can discover each other. If they are in omnidirectional mode, they can discover devices in any direction, similar to traditional frequency bands, but their communication range is very limited. Alternatively, they can extend their range by using a narrow beam, but then they need to explore a lot of directions. Existing solutions alternate between using wide and narrow beams. They also leverage the fact that the control packets used during association use a low bit rate so their range is effectively longer. Once a basestation and client can communicate, they can optimize their antenna setting to maximize range and throughput by "pointing" narrow antenna beams at each other.

For stationary clients, the client and basestation can both remember each other's position, or more precisely, they can reuse the parameter setting for the antenna array that optimized communication. Note that when a basestation is communicating with a client, it will likely be deaf to transmissions of other clients, so it will need to regularly listen to each of its clients. Mobile clients are more challenging, since both the basestation and client need to track each other by updating the settings of their antenna array.

While mmWave communication is challenging, studies have shown that it can be effective in both WiFi and cellular (Aggarwal et al., 2019; Wang et al., 2020). Unfortunately, 802.11ad (Nitsche et al., 2014) has not been very successful. It has mostly been used to create point to point links between traditional (2.4 and 5 GHz) basestations and for simple point-to-point applications. 802.11ay has addressed many of the inefficiencies of 802.11ad, so it may see broader use. For example, one of the goals was to extend the range of 802.11ay to over 100 m, compared to 10 m for 802.11ad.

### 4.3.7 Interoperability

While the release of increasingly more powerful WiFi versions is attractive to users, it raises the question how devices using different Wfi versions can communicate with each other.

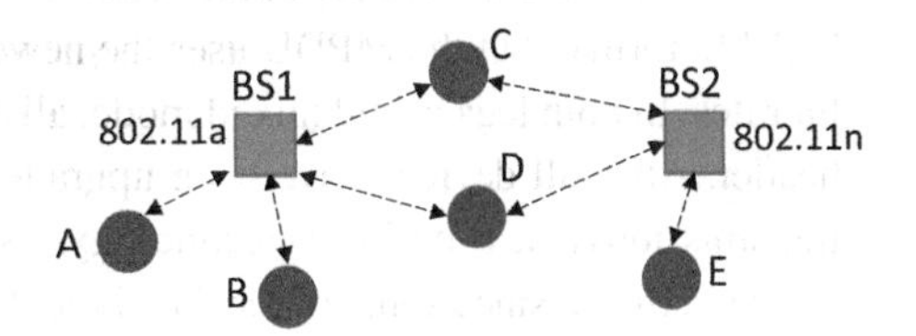


**Fig. 4.9** Example of a mixed WiFi deployment

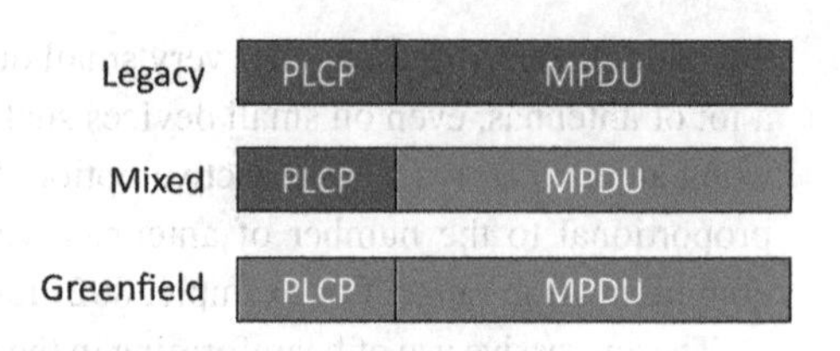


**Fig. 4.10** Interoperability 802.11a (green) and n (blue)

Figure 4.9 shows an example of mixed WiFi deployment. The dashed arrows show what clients are within range of each of the basestations. Let us first focus on AP1 and clients A-D, which use 802.11a in the 5 GHz band. When user D upgrades her device to 802.11ac, she no longer will have WiFi connectivity with either basestation. Even worse, if BS1 is upgraded to 802.11ac, no older devices will be able to use it. Clearly, this provides a strong disincentive to upgrade. The solution is easy. Chipsets for new WiFi versions always support the older versions as well, so they can communicate with legacy devices. For example, if BS1 and device D have been upgraded to 802.11ac, then BS1 will use 802.11a for communication with A-C, and 802.11ac for D.

However, mixed deployments also create a trickier problem. As discussed in Sect. 4.3.4, virtual carrier sense (VCS) is an important mechanism to prevent collisions. While chipsets for new WiFi versions can be designed so they can access VCS information used by older WiFi versions, older devices will not be able read the VCS information sent by newer WiFi versions, since they often use more advanced modulation and coding techniques. For example, while both both 802.11a and 802.11n use OFDM, their formats are not compatible. This means that in our earlier example, devices A-C will not be able to access to VCS information in packets exchanged by BS1 and 802.11ac device D.

The solution is to have newer versions of WiFi include VCS related information in a format that can be read by previous WiFi versions. It is stored in the header of the PHY Layer Convergence protocol (PLCP), which is part of the WiFi physical layer. The PLCP header is the first header in a WiFi packet and it contains bit rate and packet duration information, which are needed to demodulate and decode the packet and to update the NAV. Figure 4.10 shows at a high level how the PLCP header can support interoperability in the mixed 802.11a and 802.11n deployment in our simple example. The MAC Protocol Data Unit (MPDU) contains the WiFi header and data. 802.11n devices can use either the 802.11n (blue) or 802.11a (green) format and modulation for the PLCP and MPDU.

This results in three operating modes. First, *Legacy mode* is used for communication that involves an 802.11a device. *Mixed mode* is used when two devices using 802.11n communicate while there are 802.11a devices in the environment. The PLCP header uses the 802.11a format, but the MPDU uses the newer 802.11n modulation, which supports higher bit rates. In both legacy and mixed mode, all devices within range are able to read the PLCP header. After all devices have been upgraded, *Greenfield mode* can be used, allowing all transmissions to use the fastest technology. As expected, ensuring interoperability introduces extra overhead, since some of the data is sent at lower rates in both legacy and mixed mode.

The above example focuses on a simple case. In general, the environment may have traffic using as many as four WiFi versions in the same part of the spectrum (e.g., a, n, ac, and ax in the 5 GHz band). In addition, some basestations may use channel bonding to use more spectrum, but they must still be able to communicate with, and make VCS information available to, older devices that only support narrower frequency bands. For example, in Fig. 4.9, BS1 could use WiFi channels 54 and 56 (40 MHz) while D could just channel 54 (20 MHz). Additional extensions are needed to enable access to VCS information.

### 4.3.8 Other WiFi Features

Over the years, IEEE 802.11 working groups have defined many WiFi features related to the operation and management of WiFi. A few of these features can be of particular interest to users and researchers who use WiFi.

A first feature is 802.11h, an amendment to the 802.11 standard for "Spectrum and Transmit Power Management". It was introduced because in some countries, the 5 GHz band was already used for certain types of radar and satellite communication before it was opened up for WiFi use. WiFi basestations that detect the presence of such primary users must make sure that their communications with client devices does not interfere with the primary user. 802.11h specifies two mechanisms to do this. First, basestations can control the transmit power used in the BSS. Second, basestations can select, and move to, another channel (dynamic frequency selection), after instructing their associated devices that they should also move to the new channel. Interesting enough, both techniques are also useful for optimizing WiFi deployments. Transmit power control can be used to limit interference with nearby WiFi basestations, while channel switching can be used to move the basestation and its associated devices from a crowded channel to a channel with lower utilization.

Another WiFi feature that is directly relevant to users is Power Save Mode (PSM), which was part of the original standard. WiFi devices are often mobile so conserving battery power is important. Even when a device is not actively communicating, the WiFi radio needs to listen to all WiFi traffic in its channel to make sure it does not miss any packet addressed to it. The power spent in this idle receive state can easily dominate WiFi power consumption. PSM allows a device that is not actively communicating to put its WiFi radio into sleep mode most of the time. When a WiFi device is in PSM and it needs to send a packet it can simply wake up the radio. Receiving packets is harder since they arrive at the basestation, which cannot wake up the device's radio. Figure 4.11 illustrates the PSM protocol. The basestation keeps track of what devices are in PSM and it queues any packets it receives for the PSM devices.

WiFi basestations periodically send a beacon (described in the next section) and when it has packets queued for devices in PSM mode, it includes a Traffic Indication Map (TIM) that list all devices that have packets waiting. Devices periodically wake up to receive (possibly a subset of) the beacons and when they are listed in the TIM, they poll the basestation,

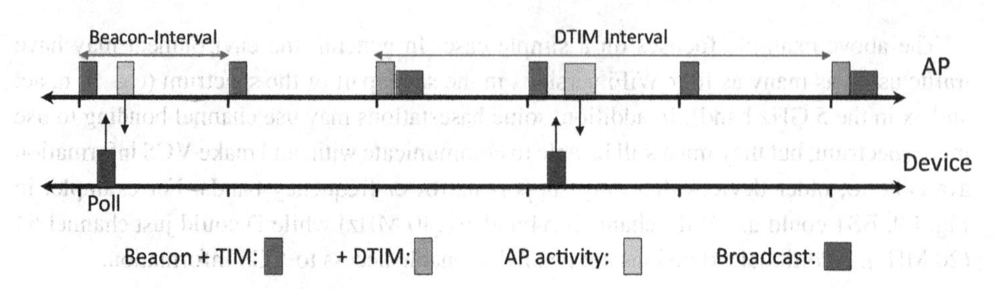


**Fig. 4.11** WiFi power save mode

requesting their packets. If the queue includes broadcast or multicast packets, it also includes a Delivery Traffic Indication Map (DTIM) in the beacon. A DTIM is include in a only a subset of the beacons, and a all devices need to wake up for those beacons, so they can receive the broadcast and multicast traffic, if there is any. Power save mode is very effective at saving battery power and it is turned on by default on WiFi devices. One obvious drawback is that there is a non-trivial delay when a packet arrives at the basestation while the destination device is in PSM.

Finally, device localization is an important capability for many mobile applications. While GPS can generally be used outdoors, it is not available indoors. The WiFi Fine Time Measurement (FTM) protocol introduced as part of 802.11mc in 2016, allows WiFi devices to measure the distance to basestations that are within range. This is done by calculating the Time of Flight (ToF) of transmissions between the device and basestation using the RTT obtained from packet exchanges between the two devices. Using triangulation, the device can then calculate its position, based on the position of the basestations, with 1–2 m accuracy (Ibrahim et al., 2018).

### 4.3.9 Connecting to a Network Using WiFi

The primary control function of WiFi is allowing clients to "connect" their WiFi device to the wired network infrastructure through a basestation. This step involves three functions: (1) associating with an basestation, (2) establishing IP connectivity, and (3) authenticating and getting authorization to access LAN and Internet services.

#### WiFi Association

Associating with an basestation also involves two steps. First, WiFi clients must select the basestation, or more precisely the service set, they would like to use. The service set can be a BSS, e.g., a home WiFi network with a single basestation, or an ESS, e.g., on a campus (Sect. 4.3.1). Service sets advertise themselves by periodically by sending *beacon frames*, typically with a frequency of one beacon every 100 ms. Beacons are broadcast packets that include the Service Set Identifier (SSID) of the service set that the basestation belongs to, in addition to information about its capabilities, such as the supported bit rates and physical layer information needed to communicate with the basestation.

WiFi devices that want to connect to a WiFi network must first discover what service sets are available. A first approach is *passive scanning*, which means that the device passively listens for WiFi beacons. This can be time consuming, considering that there are many WiFi channels and devices must listen on each channel for at least 100 ms, given that beacons are transmitted periodically. Alternatively, devices can do *active scanning* which means that they send a Probe Request, inviting basestations to send a beacon. This is faster since devices can spend less time on each channel (e.g., 10–20 ms) but it has higher overhead.

Once the WiFi device has identified a set of available service sets, it must pick the one that it wants to use. One way of doing this is to provide the user with a list of advertised SSIDs, so she can pick the one she wants to use. For example, when a user is on the CMU campus, she can pick CMU-secure. Alternative, depending on the user setting, the device can automatically pick the service set without user involvement, for example, based on past use of the service set.

Finally, both basic and extended service sets sometimes advertise multiple SSIDs, offering different types of connectivity. For example, a campus WiFi network could offer "secure" and "guest" network access, where guess access only provides access to the Internet but not to on-campus services, while "secure" access also provides connectivity to the internal campus network.

In the second step, the devices need to associate with an basestation. For basic service sets, the device will use the only available basestation, while for extended service sets, the basestation is picked automatically, for example based on signal strength. Association is done using an exchange of control packets that are used to reach an agreement on what PHY and WiFi parameters to use.

### IP Connectivity

Once a device has associated with a basestation, it can send and receive packets at layer 2 (WiFi). However, IP connectivity is needed for accessing services. Similar to wired devices, WiFi devices use the Dynamic Host Configuration Protocol (DHCP) to obtain the information needed to use IP. New devices broadcast a DHCP request on the local network and a DHCP served will respond with the necessary information. This includes an IP address for the device, the IP prefix for the local subnet (discussed in Sect. 5.1.1), and the IP address of the gateway router that connects the WiFi network to the local network infrastructure and/or the public Internet (Fig. 4.4).

### Security

Security is another important aspect WiF association since potentially anybody can try to associate with the network. Today, virtually all WiFi access links are encrypted. This is important, since it is very easy to listen in on wireless links, so sending data in the clear is a huge privacy liability. Authentication and authorization are more complicated and several alternatives are in use today.

The simplest form of authentication is MAC-based Access Control, in which network access using WiFi is controlled based on the IEEE datalink address of the device. The basestation or authentication server has a "white list" with the MAC addresses of devices that are allowed to access the network. The benefit of MAC-based access control is that it is simple and fast. Unfortunately, it is not very secure, since MAC addresses can easily be spoofed. An alternative is to use the IEEE 802.1x, which defines a protocol for authenticated and encrypted access to IEEE 802 networks. It provides access based on a password.

Today, many small WiFi networks supports encryption based on the WPA2-PSK protocol. It is based on pre-shared keys and replaces the older and less secure WPA and WEP standards. When the basestation is installed, a shared cryptographic key is configured on the basestation based on a password provided by the administrator of the network. Authorized users are given the password, which they can use to access the network over an encrypted channel to the basestation. One benefit of this solution is that it does not require individual user account since the password can be shared by all authorized users. The WPA2-PSK protocol is widely used in home networks and, more generally, in small networks with a limited number of users, e.g., hotels and hotspots.

The security of WPA2-PSK depends on the security of the password and as the number of users increases, the risk that the password is leaked grows. WPA2-Enterprise offers a more secure solution. It uses the Extensible Authentication Protocol (EAP) (Vollbrecht et al., 2004) and a RADIUS (Aboba & Calhoun, 2003; Rubens et al., 2000) server to support Authentication, Authorization, and Accounting. EAP is a standard framework that supports different types of authentication. Connecting to a WiFi network using the EAP protocol involves two steps. First, the client device must associate with a basestation in the desired service set. Once this is completed, the device is given restricted access to the LAN. Specifically, it can communicate with the RADIUS server but it cannot access other services. Second, the client authenticates with the RADIUS server to get access to more services. Restricting a device's access during authentication using the RADIUS server is typically done using a Virtual Private Network (VPN) or virtual LAN (VLAN). Finally, the RADIUS server gives the device access to network services for which it is authorized.

### 4.3.10 WiFi Deployments

WiFi deployments are very diverse, ranging from home networks with a single basestation, to small deployments with a few basestations in stores or hotels, to large campus deployments with thousands of APs. The main difference across these deployments is the degree of management that is used to coordinate across the basestations sharing the spectrum in a geographic area. In this section, we briefly discuss the two extremes of this spectrum: managed and chaotic WiFi deployments.

#### Managed Deployments

Managed WiFi deployments are larger networks that are installed and managed by a professional network management group. The general rule for unlicensed spectrum is that anybody can use it, as long as they comply with FCC regulation. Campus deployments are an exception in the sense that the organization that owns the campus has considerable control over the use of unlicensed spectrum use on campus. For example, they can control who can install basestations which is important not only for performance reasons, but also for security since rogue basestations connected to the wired infrastructure can be used to bypass security measures imposed on the WiFi network.

A key design decision in building a high-performance WiFi network is the placement of the basestations in the physical environment. Placement is primarily driven by *coverage* requirements: what parts of the space must offer WiFi access. Today, the general expectation is that WiFi access is universal, not just indoors, but also in frequently used outdoor spaces. Identifying locations for basestation placement is a labor-intensive process that involve measuring the coverage provided by the many candidate locations. Besides coverage, several other factors impact basestation placement. For example, areas that are often crowed (large auditoriums, food courts) may need to be covered by several basestations. Also, the coverage provided by a basestation depends on the frequency band (2.4 vs. 5 GHz). Finally, once mmWave WiFi sees wider adoption, its heavy reliance on beamforming will add another dimension to the basestation placement problem.

Once basestations have been installed, the most critical decision in optimizing performance is how to assign frequency bands to each basestation. This must be done so that frequency bands are not reused in cells that are too close together, since it can result in interference and hidden and exposed terminal scenarios (Sect. 4.3.4). A simple example using three frequency bands (shown as blue, green, and red) is shown in Fig. 4.12. Besides frequency selection, transmit power control can also be used to manage interference. In the simplest case, frequencies can be statically assigned to basestations. However, it can be more efficient to assign frequencies dynamically based on load, for example increasing the coverage of a large auditorium depending on whether it is used. Some basestations can even be switch off if their capacity is not needed, for example, at night.

Campus networks, including both wired and wireless subnets, are often centrally managed in an integrated fashion. They rely heavily on virtualization to isolate subnets with different requirements, such as different WiFi service sets, using for example VLAN technology. For WiFi specifically, the centralized controller decides on key decisions such as the allocation of frequency bands, width of frequency bands, transmit power, and assignment of nodes to basestations. Decisions are based on information, such as traffic load and interference levels,

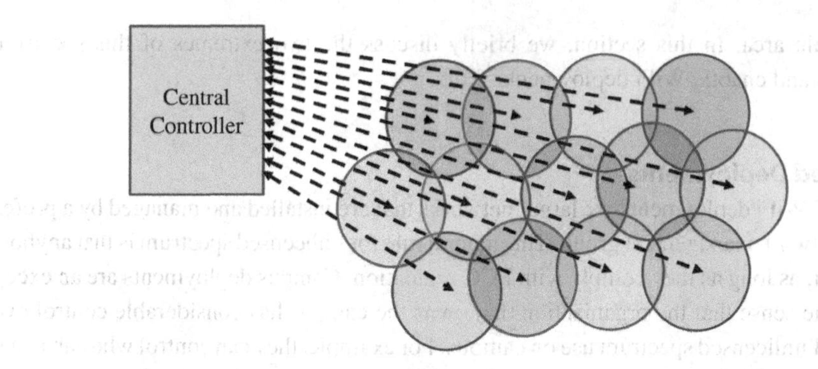


**Fig. 4.12** Simple centrally managed WiFi deployment

which is collected by the basestations. Basestations are relatively dumb devices that simply forward packets based on the configuration specified by the centralized controller.

### Chaotic WiFi Deployments

At the other end of the spectrum, we have geographic areas such as residential neighborhoods or apartment buildings, where many basestations that part of small services sets share the spectrum. Examples are home network and coffee shops. They differ from managed networks in two ways. First, the individual service sets are often unmanaged: users buy a basestation, connect it, and use it. Second, there is no global management of the basestations sharing the spectrum: they each operate independently. For these reasons, these deployments are sometimes referred to as *chaotic* (Akella et al., 2005).

Most of the optimizations used in managed deployments cannot be applied in chaotic deployments. Basestations are placed in locations that are convenient, without considering coverage and there is no coordinated configuration of the basestations. In the early days of WiFi, the situation was especially bad. For example, manufacturers of basestations would configure their basestations with a default setting, e.g., channel 1 in the 2.4 GHz band, so it was common to see many basestations in the same area use the same channel, causing unnecessary contention. Fortunately, today's basestations are much smarter and are to some degree self-configuring. For example, they regularly scan the available frequency bands to find a band that is has low utilization, which helps in distributing traffic across the channels.

In between these two extremes, many areas contain both smaller managed networks and single basestation service sets that are unmanaged. Smaller managed networks can use many of the optimizations discussed above, but since they do not control all devices in the areas, they are less effective than campus deployments.

### 4.3.11 Mobility

Since the transmission range of WiFi is limited, typically 50–100m or less, mobile users will switch between basestations as they move around. We will consider two scenarios. First, the user has to switch between basestations that are part of different service sets. Associating with a new service set is usually quite slow, even when no manual user intervention is needed (Bychkovsky et al., 2017a; Pei et al., 2006), so ongoing transmissions are likely to be disrupted.

The second scenario is a user who switches between basestations that are part of the same service set. Originally, WiFi did not provide support for intra-service set roaming, since mobility was not a priority. Later on, a number of extensions were introduced to speed up roaming, including, 802.11k, 802.11v, and 802.11r. They provide features that can reduce the overhead of some of the most time-consuming operations used in association. Examples include providing devices Neighbor Reports with information on neighboring basestations, that can be used to optimize scanning, informing devices when it might be advantageous to switch basestation, and speeding up authentication. These features must be supported both by the software (OS and WiFi device driver) of the wireless device and by the service set that the user is connected to before they can be used.

### 4.3.12 WiFi for Special Environments

A number of WiFi standard for specialized environments have been defined.

A first example is 802.11ah, known as WiFi HaLow which operates in license-exempt frequency bands below 1 GHz, such as the 900 MHz ISM band (Sun et al., 2013). 802.11ah is designed for IoT devices by offering longer transmission ranges with lower energy consumption than traditional WiFi operating in the 2.4 and 5 GHz frequency bands. It supports transmit rates up to 347 Mbps. The protocol was approved in September 2016 and published in May 2017.

Another example is 802.11af (Flores et al., 2013), which supports WiFi in the TV white spaces (Sect. 2.1.4). 802.11af is an extension of the WiFi protocol, which was defined for use in unlicensed spectrum. Specifically, it is modified to comply with the regulatory requirements specified by the FCC, by using a geolocation database to avoid interference with primary users.

## 4.4 Cellular: High Performance in Licensed Spectrum

We now discuss the network technology used in cellular networks. While WiFi and cellular have over time used similar PHY technology, their protocols are very different since WiFi is designed for the licensed spectrum, while cellular operates in licensed spectrum.

### 4.4.1 Cellular Versus WiFi

Cellular service is an alternative to WiFi for high performance data service, especially for mobile users. Cellular differs from WiFi in three significant ways: spectrum use, service model, and protocols.

First, cellular service uses *licensed spectrum*, while WiFi operates in the unlicensed spectrum. In order to provide service, cellular operators need to acquire spectrum from the FCC. Because of the high demand for cellular service, the FCC regularly frees up more spectrum for "mobile broadband service". This spectrum is then made available through auctions. Acquiring spectrum is expensive. For example, a recent spectrum auction raised $80 billion for 5,684 licenses covering different areas in the US (https://docs.fcc.gov/public/attachments/DOC-370267A1.pdf). The use of licensed spectrum has a significant impact on both the type of service that cellular operators provide compared with WiFi, and the technology that is used to provide it.

Second, while WiFi service is often free, cellular operators always *charge users for service*. One reason is the high cost of acquiring spectrum licenses, while the use of the unlicensed spectrum is free. Another reason is that the cellular infrastructure is much more complex than WiFi and it is expensive to build, maintain and operate. Note that users sometimes have to pay for WiFi service, e.g., in hotspots and hotels). This payment is used to cover the cost of the WiFi infrastructure and Internet access, not spectrum use. Considering the cost of cellular service, users expect "good" service. More formally, this takes the form of a Service Level Agreement (SLA) between the user and the provider that specifies the provider's commitments for voice and data service. Voice is subject to stringent performance requirements, while the data service quality is more loosely defined (e.g., data volume and expected throughputs). In contrast, WiFi service is always best effort.

Finally, considering the need to provide good and consistent service to customers, cellular networks use *scheduled access* for medium access control, in contrast to WiFi, which uses random access (see Sect. 4.2.5).

### 4.4.2 Origins of Cellular Technology

The first "wireless telephone service" was developed at AT&T Bell Labs in the mid-1940s and it became commercially available in some US cities in 1946. It consisted of a transmitter on a high building providing telephone service across the city with just a few channels. Figure 4.13 shows how multiple "transmitter and receiver" towers communicate with "receivers". As we can see in the figure, the first "receivers" of the cellular service were vehicles, which were equipped with a car phone. At that time, the technology to build a cellphone that could be carried around easily, let alone fit in a pocket, did not exist. The development of the transistor and improvements in battery technology were needed to build more traditional cellphones.

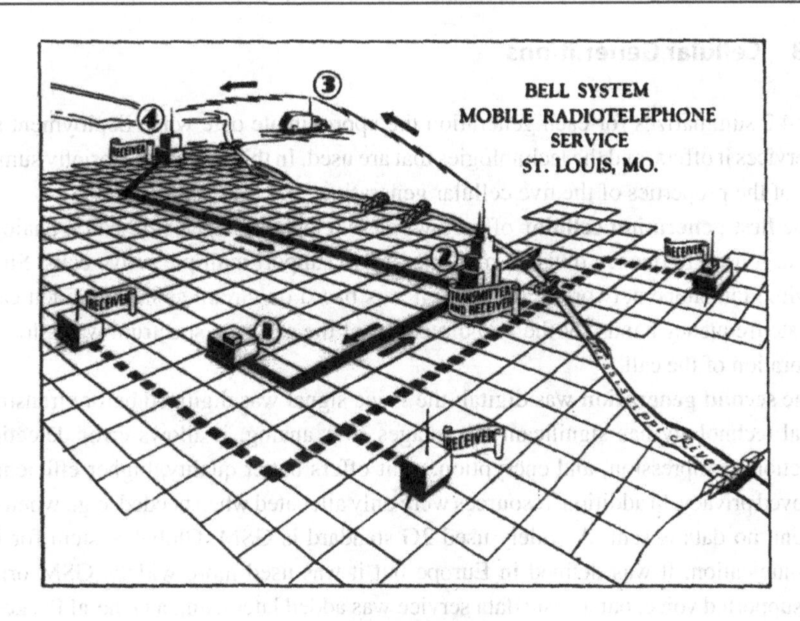


**Fig. 4.13** Concept of first "Mobile Radio" telephone service

The concept of a *cellular* telephone network was also developed at Bell Labs shortly after that. A description of the concept can be found in an internal Bell Labs memorandum written by Donald Ringer in December 1947 (https://web.archive.org/web/20120207062016/http://www.privateline.com/archive/Ringcellreport1947.pdf). The document shows how a city can be partitioned in smaller areas (cells, represented by circles), each with their own tower. Widespread deployment of cellular telephone service had to wait until after 1968, when the FCC made spectrum available for this new wireless service.

The evolution of cellular technologies is structured as a sequence of *generations*, starting with 1G in the 1980s. Each generation consists of four steps. the first step is specifying requirements based on technology and market projections. The next steps are standardization, product development and commercial deployment. The time interval between generations is about 8–10 years. Today, 5G is being deployed, while the standardization efforts for 6G are starting. Since 3G, the definition of cellular standards has primarily been the responsibility of 3GPP and the ITU. The Third Generation Partnership (3GPP) is a group of seven standards organizations representing stakeholders in different parts of the world. They are heavily involved in both the definition of requirements and standards. The International Telecommunications Union is responsible for coordinating the spectrum use internationally, but it is also involved in the definition of standards and regulations. For 1G and 2G, competing standards were used in different regions in the world. These standards did not interoperate.

### 4.4.3 Cellular Generations

Table 4.2 summarizes for each generation the approximate date when deployment started, the services it offers, and the technologies that are used. In this section, we briefly summarize some of the properties of the five cellular generations.

The **first generation cellular** offered wireless telephone service based on analog modulation. It used frequency division multiplexing to support multiple active calls. Similar to the wired telephone network at the time, it was based on circuit switching Each call used separate frequency bands for the two directions of the call, and spectrum was allocated for the duration of the call.

The **second generation** was digital: the voice signal was digitized before transmission. Digital technology has significant advantages over analog: it allows error detection and correction, compression, and encryption, so it offers better quality, higher efficiency, and improved privacy. In addition, resources were only allocated when needed, e.g., when a caller is silent, no data is sent. A widely used 2G standard is GSM, Global System for Mobile Communication. It was defined in Europe but it was used more widely. GSM originally only supported voice, but a basic data service was added later using a General Packet Radio Service (GPRS). Data was supported by using some of the resources intended for voice, i.e., data was layered on top of a voice-oriented network. GPRS only supported simple applications such as text messaging.

The **third generation** included many standards, several of which used spread spectrum to improve the robustness of the transmission in the presence of frequency-selective attenuation and fading. Most of the 3G standards provided direct support for both voice and data.

Today, we mostly use **fourth generation** technology, as supported by Long Term Evolution (LTE). In contrast to earlier generations, LTE implements a data-only service based on IP, and Voice is layered on top using Voice over IP technology. The LTE physical layer is based on OFDM, the same technology used by all recent versions of WiFi. The **fifth generation** uses similar technologies but it also uses the millimeter wave band and it targets a broader set of application domains.

**Table 4.2** Cellular technology generations

| Version | 1G | 2G | 3G | 4G | 5G |
|---|---|---|---|---|---|
| Deployment | 1982 | 1991 | 1999 | 2008 | 2018 |
| Services | Analog Voice | Digital Voice + GPSR (Gen 2.5) | Digital Data + Voice | IP based + VoIP | IP based + VoIP |
| Technology | FDMA | FDMA + TDMA | CDMA + other | OFDMA + SC-FDMA | OFDMA + SC-FDMA |

Next, we describe LTE in more detail in Sects. 4.4.4, 4.4.5, and 4.4.6. We then discuss an improved version of LTE, called LTE Advanced, and initial 5G technologies in Sects. 4.4.7 and 4.4.8.

### 4.4.4 Cellular Network Architecture

Figure 4.14 shows the high level architecture of an LTE network covering a geographic area such as a metropolitan region. It consists of two subsystems. First, the Radio Area Network (RAN), formally known as the Evolved UMTS Terrestrial Radio Access Network (E-UTRAN), supports wireless communication between mobile devices (called UE for User Equipment in the figure) and basestations mounted on cell towers (called eNodeB for Evolved NodeB). The second subsystem, the Core Network, formally the Evolved Packet Core, is responsible for the control and management of the cellular network and for connectivity to the Internet. The combination of the EPC and associated RAN in a geographic region is called an Evolved Packet System (EPS). Previous cellular generations had a similar architecture, although some functions were sometimes implemented in a different system module.

The two subsystems communicate using two S1 interfaces, one for data (S1-U) and one for control (S1-MME). These two interfaces are designed to allow the two subsystems to evolve independently. This is important since they support different functions and are enabled by different technologies. The goal of the RAN is efficient spectrum use, which depends largely on improvements in radio technology. In contrast, the core network deals with traffic management and it is effectively a data center, benefiting from enhanced computing hardware and software. There is also a standard X2 interface for coordination between eNodeBs in the same RAN.

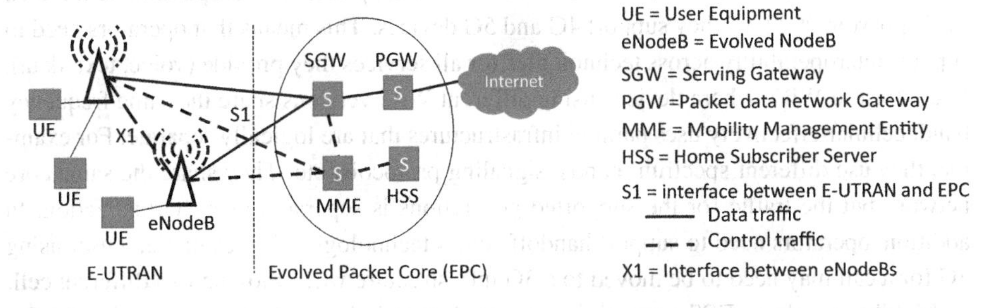


**Fig. 4.14** LTE network architecture

### Component Responsibilities

eNodeBs are responsible for managing wireless connectivity with the mobile devices. They also manage handoffs between eNodeBs in the same RAN using the X2 interface. Handoff between basestations in different RANs is managed by the core networks.

In the packet core, the Serving Gateway (SGW) is responsible for managing the traffic between the eNodeBs and the packet core. As mentioned earlier, LTE is IP based so all traffic is sent as IP packets. The S1-U protocols and packet core support Quality of Service for voice and other services. It relies on virtualization to isolate traffic in different classes, e.g., voice versus data, in addition to isolating traffic from different cellular generations, e.g., 3G and 4G. The Packet Data Network Gateway (PGW) connects the core network to the Internet.

The Mobility Management Entity (MME) implements all control functions related to subscribers and their communication sessions. It is responsible for authentication users when they connect to the network. It also tracks users when their phones are idle, so the core can establish a communication session when there is an incoming call or text message. In addition, it manages all communication session between the mobile devices in the RAN and the SGW. The Policy and Charging Resource Function (PCRF) provides Quality of Service (QoS) coordination. For example, it makes sure that communication session have performance properties consistent with the user's profile. The Home Subscriber Server (HSS) maintains a database with information of all subscribers who are associated with this core network.

### Control Plane

The control plane functions of a 4G cellular network are, by themself, very complicated. In addition, cellular operators must support multiple cellular generations to provide connectivity for customers with older mobile devices. For example, most US operators terminated 3G support in 2022, so they support 4G and 5G devices. This means that operators need to support interoperability across technologies for all services they provide (voice, text, data). In contrast to WiFi, where devices using different WiFi versions share the same frequency band, cellular effectively uses parallel infrastructures that are logically separate. For example, they use different spectrum bands, signaling protocols, etc. They share the same core network but the traffic for the supported generations is separated using virtualization. In addition, operators need to support handoff across technologies. For example, a user using 4G for a call may need to be moved to a 3G infrastructure when moving to a different cell.

Mobility inside an EPS network is supported directly by the basestations using the X1 Interface. Similar to 2G and 3G, LTE also supports roaming, so approved subscribers from other EPS networks can use the network. While visiting, subscribers receive a local identity and IP address, so the EPC modules described above can treat the visiting subscribers the same was as "local" subscribers. Note that authentication of the visiting node requires the involvement of the subscriber's HSS in its home EPS, since it stores the subscriber and authentication information. In addition, the visiting subscriber's home HSS needs to keep track of which EPC the subscriber is currently connected to so it can forward calls and text messages.

### 4.4.5 LTE PHY and Resource Management

The LTE physical layer uses OFDM and MIMO. As discussed in Sect. 3.6, OFDM supports fine grain optimization in the presence of frequency-selective attenuation and fading. This is important to get high spectrum efficiency for cellular since the long distances between cellular basestations and devices (up to a few km) can result in significant multi-path effects with large path length differences. In addition, it offers diversity in time and space which is important for dealing with fading.

LTE uses different radio access technologies for the uplink and downlink. For the downlink it uses OFDM Multiple Access (OFDMA). Figure 4.15 contrasts OFDMA with OFDM. Figure 4.15a shows an OFDM channel that is shared by multiple users using TDMA: all subcarriers during each timeslot are used by the same transmitter. In contrast, Fig. 4.15b shows that with OFDMA, the basestation can use a subsets of subcarriers to transmit data to different devices during each timeslot, i.e., OFDMA combines TDMA with FDMA. OFDMA allows more fine grain control over resources compared with OFDM, e.g., a basestation can send traffic to multiple devices at the same time.

Unfortunately, OFDM, and thus OFDMA, has a high peak-to-average power ratio, which results in high energy consumption. This means that OFDMA is not an attractive solution for the LTE uplink, since battery lifetime is an important metric for mobile devices. Instead, LTE uses a technology called Single Carrier FDMA (SC-FDMA) for the uplink. SC-FDMA has benefits similar to OFDMA, but it has a lower peak-to-average power ratio. This is achieved by adding pre and post processing steps to the OFDM standard processing pipeline.

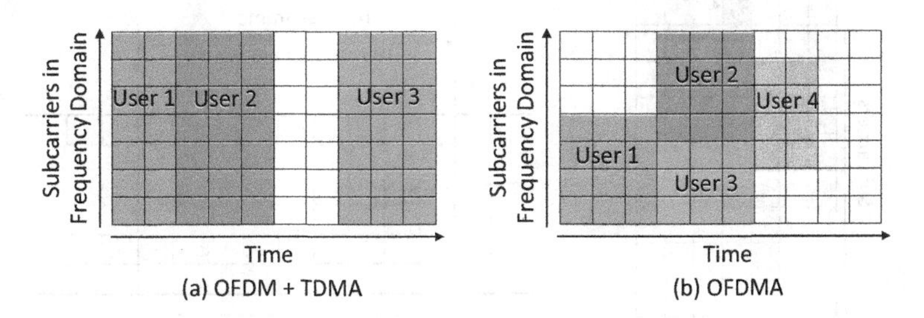


**Fig. 4.15** Comparison of **a** OFDM and **b** OFDMA

### 4.4.6 Resource Allocation

The bandwidth (Mbps) allocated to users is directly linked to their transmission time and how much spectrum they use. In this section, we briefly describe how uplink and downlink traffic are managed and we then elaborated on how resources are allocated to users.

LTE bands can be up to 20 MHz, but using a feature called Carrier Aggregation, the spectrum does not have to be contiguous. This gives operators significant flexibility in how they manage spectrum. A basestation can also support multiple frequency bands, potentially using different technologies. Cellular operators have two options for allocating spectrum resources for uplink and downlink traffic. First, with Frequency Division Duplexing different frequency bands are used for uplink and downlink traffic. An alternative is a technique called Time Division Duplexing, where one frequency band is used for both uplink and downlink traffic, alternating in time. The tradeoffs between these techniques are complicated since they differ in the complexity of the radio, spectrum efficiency, and flexibility of bandwidth allocation.

Resources are fully controlled by the basestation and they are only allocated when there is data queued up for transmission. Figure 4.16 shows how LTE manages resources in the downlink spectrum band based on OFDMA. It shows 15 KHz subcarriers each transmitting a stream of symbols corresponding to the vertical and horizontal dimension respectively. They form a grid of *resource elements*. Resource allocation is done on the basis of *resource blocks*, each consisting of the resource elements of 12 subcarriers and 7 symbols (Fig. 4.16a). The resources in a spectrum band are organized using resource blocks as shown Fig. 4.16b. For the downlink, the basestation can allocate resource blocks to active devices depending on the amount of queued data and QoS requirements. It also informs devices which resource blocks are used for their data.

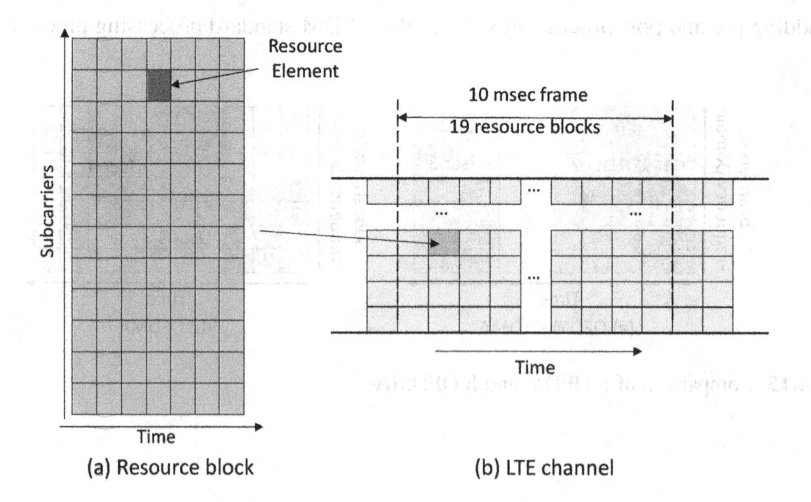


**Fig. 4.16** LTE resource allocation

The uplink uses SC-FDMA and resource allocation is also controlled by the basestation. However, the basestation does not have direct access to the status of the send queues on the devices. When a device has packets queued up, it needs to request resources from the basestation using control slots assigned to each device. The basestation then inform the device which uplink resources it can use. This step must be repeated for each packet, which needless to say, incurs a delay, as documented in several papers (Tan et al., 2021; Zhang et al., 2019).

### 4.4.7 LTE Advanced

In 2009, several improvements across the LTE protocol stack were introduced, known as LTE advanced. LTE Advanced supports wider frequency bands (up to 100 MHz) and defines better PHY technology offering higher throughputs. It also reduces latency in the control plane (e.g., latency to connect or switching to active status) and data plane (resource scheduling).

LTE Advanced also introduces small cell sizes. Traditional macro cells can have a range of a few kilometer. This large size makes it hard to optimize capacity across areas with different traffic densities and it also limits frequency reuse across cells, which is important when optimizing spectrum efficiency. In contrast, micro or picocells can have a range of 10–100 s of meters. Smaller cells make it possible to more effectively cover areas with high demand or areas that are difficult to cover using macro cells, for example because of geography (e.g., valleys) or man-made obstacles. Small cells are also one of the enablers for edge computing (Chap. 6), since they can provide mobile devices cellular access to nearby cloudlets or edge clouds with very low latency, which is critical for latency sensitive applications such as AR/VR, autonomous driving, etc.

### 4.4.8 5G

We now discuss 5G, focusing on differences relative to 4G.

#### 5G and Motivation and Focus

Figure 4.17 shows how the ITU's vision for 5G includes support for groups of applications. At the top, we have Enhanced Mobile Broadband, which corresponds to the traditional applications supported by earlier cellular generations such as web access and video streaming, but with a higher performance. Massive IoT, shown in the left corner, will support IoT deployments with very large numbers of IoT devices, each with low bandwidth requirements. These devices must be power efficient, so a major challenge is how to simplify cellular protocols that were optimized for performance to efficiently support devices that may only send a few hundred bytes every minute. While several cellular operators already support standards

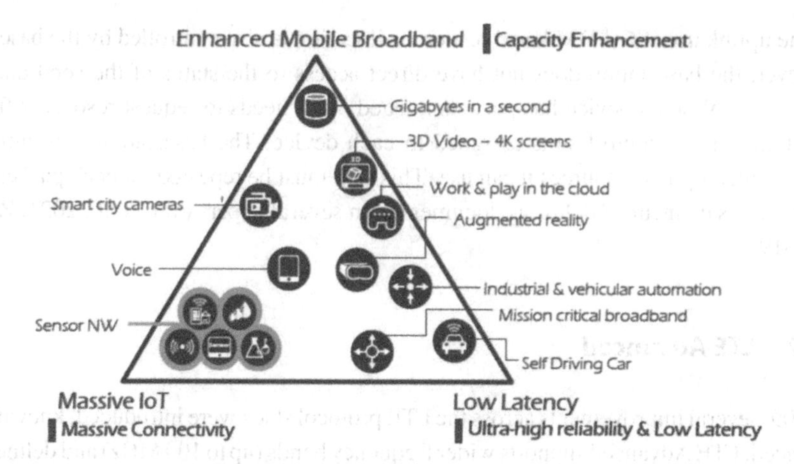


**Fig. 4.17** Application domains targeted by 5G (Figure 2 from https://www.itu.int/dms_pubrec/itu-r/rec/m/R-REC-M.2083-0-201509-I!!PDF-E.pdf. Reused with permission of the ITU)

for IoT (e.g., LTE-M, NB-IoT), they have seen only limited deployment. Finally, the right corner represents applications such as autonomous driving and industrial automation, i.e., applications that require high reliability and low latency. The last two application classes effectively expand the market for cellular operators by potentially significantly increasing the number of cellular devices.

To meet the requirements of these new application domains requires significant improvements in many performance metrics, as shown in Fig. 4.18. IMT-advanced roughly corresponds to LTE advanced while IMT-2020 corresponds to 5G.[1] For example, the figure shows significant improvements (factor of 10 or more) in both peak and average data rates as well as reductions in latency, all directly relevant to users. Other metrics such as energy and spectrum efficiency are important for operators. The mobility requirements are based on the (projected) speed of high-speed trains.

## Meeting 5G Performance Goals

The 5G requirements are very ambitious and will require improvements across the cellular infrastructure. A first step is to reuse traditional ideas: (1) allocate more spectrum for cellular and (2) improving spectrum efficiency through optimizations in both the PHY layer and the protocol stack. For example, the FCC has freed up more spectrum for 5G in the traditional bands used by cellular, namely the low (below 2 GHz) and mid bands (below 6 GHz). Spectrum was allocated for cellular operators in the mmWave band (above 26 Ghz), e.g., 25–39 GHz. As discussed for WiFi (Sect. 4.3.5), the mmWave band is attractive because

[1] IMT (International Mobile Telecommunications) requirements are defined by the ITU, while 3GPP defines LTE requirements and standards.

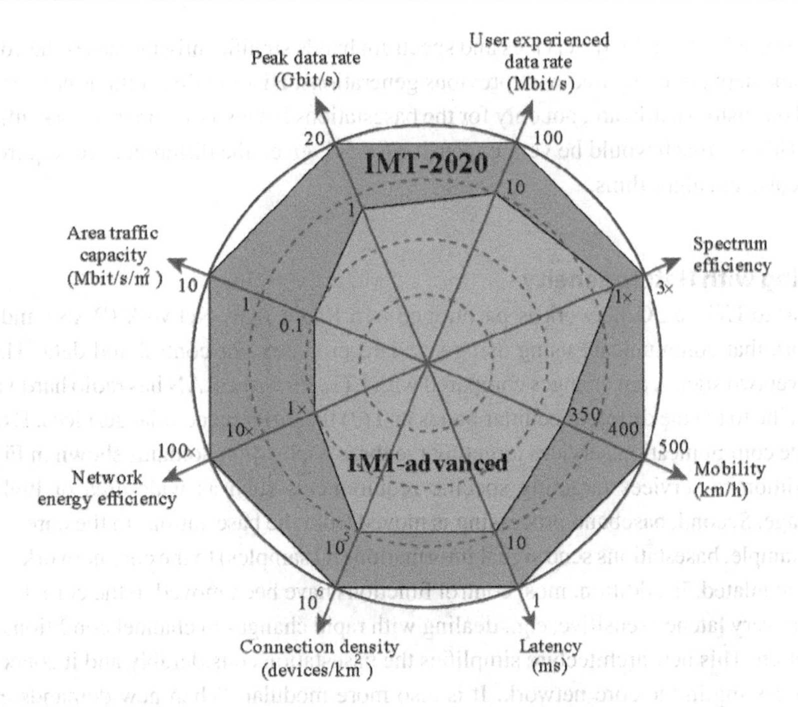


**Fig. 4.18** 5G performance goals (Figure 3 from https://www.itu.int/dms_pubrec/itu-r/rec/m/R-REC-M.2083-0-201509-I!!PDF-E.pdf. Reused with permission of the ITU)

there is a lot of spectrum available, although high attenuation makes it more challenging to use. In addition, 5G uses more enhanced coding and modulation techniques, and it also specifies a large number of antennas for fine grain beam forming and MU-MIMO. While these improvements are important, they are not sufficient to realize the 5G vision for several reasons.

First, the cellular services envisioned by 5G are very diverse as shown in Fig. 4.17. More importantly, there are conflicting requirements between the services. For example, IoT devices must be power efficient and cheap, while while mobile broadband must support high data rates. These requirements conflict since high data rates require broadband signals and aggressive use OFDM and MIMO, all of which are computationally intensive. This technology is not suitable and not needed for IoT devices. While it is possible for the two services to share the same infrastructure (e.g., cell towers, core network), they cannot use the same protocols, hardware, and software.

Second, the frequency bands range all the way from bands below 2 GHz to bands above 26 GHz. Radios in these diverse bands require different antennas (Sect. 3.7) and different techniques for optimizing performance. For example, mmWave radios rely on aggressive beam forming which requires large numbers of antennas, which is not needed or even feasible in the lower frequency bands (Sect. 3.8).

Third, 5G's diversity in services and spectrum bands significantly increases the complexity of the network compared with previous generations. Historically, cellular networks have relied on custom hardware, not only for the basestations but also for core network infrastructure. This approach would be very expensive for 5G, given the differences in requirements, protocols, and algorithms.

### Dealing with Heterogeneity

Similar to LTE, a 5G network is partitioned in a Radio Area Network (RAN) and a core network that communicate using using standard protocols for control and data. There are however two significant changes compared with LTE. First, the RAN has radio hardware that is specific to (1) the different cellular bands and (2) the different cellular services. Examples include communication services targeting the three application domains shown in Fig. 4.17, in addition to services targeting specific requirements such as wide area or high-speed coverage. Second, baseband processing is moved from the basestations to the core network. For example, basestations send signal information (IQ samples) to the core network, where it is demodulated. In addition, most control functions have been moved to the core. Functions that are very latency sensitive, e.g., dealing with rapid changes in channel conditions, are an exception. This new architecture simplifies the basestation considerably and it concentrates all processing in the core network. It is also more modular. When new demands emerge, new frontends and antennas can be added to the basestation and the software in the core network can be updated to handle the new service requirements.

### Using Commodity Technology

While earlier cellular generations relied on custom hardware in the core network, 5G uses cloud technology and commodity servers, referred to as a Cloud RAN or C-RAN., This approach has several advantages.

First, commodity servers should provide significant cost benefits compared with custom hardware. In addition, it is more flexible. For example, changes in the balance of traffic between the different service classes can be addressed by reallocating servers. Similarly, servers can be added or repurposed when services are added or change. A final benefit is that operators can use infrastructure provided by commercial cloud providers instead of having to acquire and manage their own infrastructure. Note that cellular workloads are different from those of traditional cloud users. For example, baseband processing for example is optimized based on Channel State Information (CSI) that captures conditions. It is measured by the basestations and sent to the core network. Since the CSI changes continuously, baseband processing is much more latency sensitive than traditional cloud workloads such as web services or video streaming, so cloud management techniques have to be customized.

Second, Software Defined Network (SDN) technology can be used to manage the cloud network. In SDN, the network is managed using a centralized SDN controller that supports network management applications such as routing and traffic engineering. Switches in the network are "dumb" in the sense that they do not have a control function. Instead, they are configured based on instructions from the SDN controller. SDN allows much more fine grain control over how traffic flows through the network, compared with traditional distributed protocols, such as the OSPF routing protocol. Flexible and efficient traffic control is important in cellular network, for example to support cellular services that have different requirements (Foukas et al., 2016; Riggio et al., 2015). The same is true for network slicing, a technique that partitions a single physical network infrastructure in slices that isolate different traffic classes (Foukas et al., 2017).

Finally, cellular operators rely heavily on *middleboxes* to reduce traffic on the wireless link and to optimize user experience. The Internet has traditionally used custom hardware for middleboxes but this has changed. Network Function Virtualization (NFV) technology is instead used to support middlebox functions on commodity servers. In the Internet, NVF is used both at the edge and in the cloud, so this technology can also be deployed in the C-RAN.

### Optimizing Frequency Reuse

Moving baseband processing for all basestations to the cloud has another very important, but less obvious, benefit, namely more aggressive frequency reuse across basestations. When radio processing is done at each basestation, it only has access to the CSI for the channels it is using. It can also measure the interference generated by nearby basestations and their clients using the same frequency band. However, it does not know how much interference is it generating in those cells, although basestations could potentially exchange this type of information directly. However, when baseband processing is moved to the cloud, all this information is centralized and readily available to optimize spectrum use and radio processing. Even better, the cloud has access to the transmitted and received signals. This allows the C-RAN to minimize interference and its impact on performance between nearby cells using the same frequencies. Ideally, frequencies can be aggressively reused by basestations with minimal interference, significantly improving spectrum efficiency.

# IP and TCP

# 5

## 5.1 IP and Mobility

Wireless access links do not impact IP's ability to deliver packets, but they can impact the network performance of mobile devices. After a basic introduction of the IP protocol, we discuss how we can support Internet connectivity for nomadic users.

### 5.1.1 Internet Protocol Basics

Before we describe how mobility can impact the operation of IP, we briefly describe how IP forwards packet forwarding based on IP addresses. At a high level, IP addresses have two parts, as shown at the top in Fig. 5.1. The high order bits represent a *Network ID*, sometimes also called the *prefix*, that identifies the network of the destination device (shown in blue). The remaining bits are a Node ID identifying the host in the destination network (shown in green). The sizes of the two IDs can be different, depending on the size of the networks. Today, two types of IP addresses are used in the Internet: IPv4 with 32 bit addresses and IPv6 with 128 bit addresses.

Forwarding of IP packets from a sender to a receiver happens in two phases. First, inside the network of the sender and inside each ISP on the path, the packet is purely forwarded based on Network ID, as shown by the blue arrow. The Node ID is ignored since it is not relevant until the destination network is reached. During the second phase, the packet is forwarded inside the destination network based on the Node ID, as shown by the green arrow. This hierarchical forwarding model is similar to, for example, that used for mail delivery in the postal system. Letters are delivered entirely based on the address of the residence of the intended recipient. The name of recipient is only used once the letter has been placed in the mailbox at the residence. This hierarchical addressing approach is more scalable than



P. Steenkiste, *Introduction to Wireless Networking and Its Impact on Applications*,
Synthesis Lectures on Mobile & Pervasive Computing,
https://doi.org/10.1007/978-3-031-27466-4_5

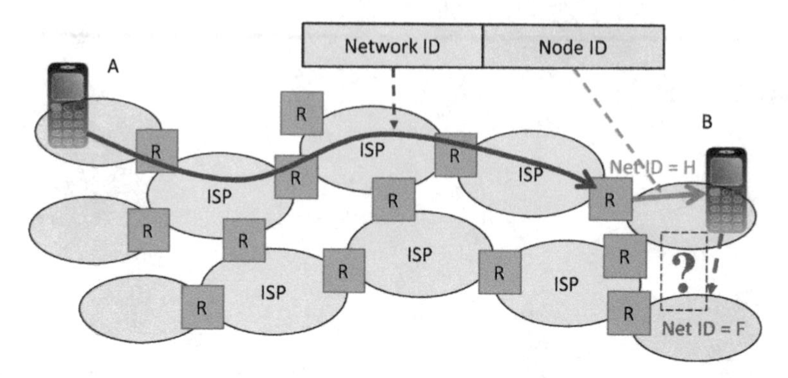


**Fig. 5.1** IP packet forwarding based on network and node IDs (Mobile phone figure from https://freesvg.org/img/silverstreak_Mobile_Phone.png. Reused under Creative Commons license CC0)

using flat addresses since routers have to store a much smaller number of addresses in their forwarding tables. Unfortunately, how IP forwards packets can be a problem when a user is mobile, as is shown at the right in Fig. 5.1. When a user moves to a different network while keeping the same IP address, packets sent to the user's mobile device will be delivered to the wrong network. In this section, we discuss the problem and possible solutions for two scenarios.

### 5.1.2 Internet Access for Nomadic Users

Nomadic users use their mobile device in different locations. However, they do not use it will they are traveling from one location to another so no network connectivity is needed while the user is mobile.

#### WiFi: Mobile IP

Supporting nomadic users became important in the 90's when laptops started to be used widely thanks to the availability of WiFi. At that time, IP addresses were assigned to computers *statically*, which meant that when nomadic users used their device in a network other than their home network, called a "foreign" network, packets sent to them would be delivered to the wrong network. Specifically, as shown in Fig. 5.1, after moving (red arrow), packet would continue to be delivered to the home network (with ID H) while the mobile host is in a foreign network (with ID F).

An early solution to this problem was the mobile IP protocol (Perkins, 1997), which was standardized by the IETF (Perkins, 1996). Figure 5.2 illustrates the key idea of Mobile IP. Assume we have a mobile host (dark blue circle) visiting a Foreign Network. When a "corresponding host" (light blue circle) sends a packet (arrow 1) to the mobile host, the

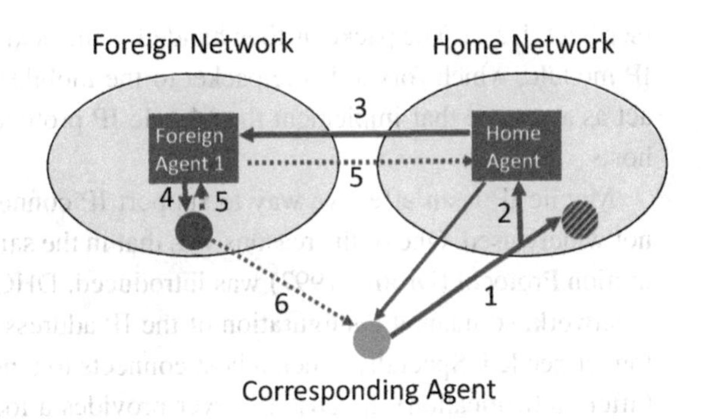


**Fig. 5.2** Operation of the mobile IP protocol

packet is delivered to its Home Network. Since the mobile host is absent (blue dashed circle), the packet is instead delivered to a Home Agent (arrow 2) who is responsible for forwarding it to the mobile host. This is done by sending it (arrow 3) to the Foreign Agent of the Foreign Network where the mobile host is currently connected. The Foreign Agent then forwards the packet to the mobile host (Arrow 4). Packets sent from the mobile host to the corresponding host can either follow the reverse path (Arrow 5), or they can be sent directly over the Internet (Arrow 6).

Forwarding of packets between the Home and Foreign Agents is done using *IP encapsulation*, which is widely used in the Internet. The concept is illustrated in Fig. 5.3. Figure 5.3a shows an IP packet sent between the two communicating devices, so the source and destination IP addresses correspond to those of the corresponding and mobile hosts. To forward such IP packets from the home to the foreign network, the Home Agent adds an additional "outer" IP headers, which uses the IP address of the foreign agent as the destination, as shown in Fig. 5.3b. The network forwards the packet based on the outer IP header while ignoring the original IP Header, which is treated as payload. When the packet arrives at the foreign agent, the outer IP header is removed (since the packet has reached its destination) so the use of encapsulation is transparent to the higher layer protocols and the application

| Original IP Header | IP Payload |
| --- | --- |

(a) Packets exchanged between Corresponding and Mobile Host

| Outer IP header | Original IP Header | IP Payload |
| --- | --- | --- |

(b) Packets exchanged between Home and Foreign Agents

**Fig. 5.3** IP-in-IP encapsulation

(see Sect. 1.2.1). The packet is then handed to the next higher protocol, in this case again the IP module, which forwards the packet to the mobile host. The Home and Foreign Agents act as a service that implement the Mobile IP protocol to support connectivity for mobile hosts.

Mobile IP is an effective way to support IP connectivity for nomadic users, but it was not widely used. One of the reasons was that in the same period, the Dynamic Host Configuration Protocol (Droms, 1997) was introduced. DHCP automates how devices connect to a network, so manual configuration of the IP address and other network information is no longer needed. Specially, when a host connects to a network, it sends a DHCP request and (after authentication) the DHCP server provides a local IP address and other information needed to use the network. With DHCP, devices of nomadic users will receive an IP address associated with the foreign network, so the corresponding and mobile host can communicate directly.

This still leaves the question of how a corresponding host can obtain the IP address of a mobile device when it is not in its home network, i.e., when sending the first packet of a TCP session. The challenge is that the corresponding host does not know the "foreign" IP address of the mobile host. Fortunately, in practice this is not a problem. Most Internet traffic is client-server traffic, in which the client initiates the connection to a server. It would be very unusual to run a server on a mobile device, since servers typically have to be available all the time. This means that for mobile device, the first packet of a connection is typically sent by the mobile device, which indirectly provides the server with the current IP address of the device. Authentication of a client to the server should be done at the application level, since IP addresses are not a reliable way of authenticating users or their devices.

### Virtual Private Networks

There are use cases in which a nomadic user needs to use an IP address associated with a remote network. For example, it is common for organizations to limit access to sensitive internal services (e.g., dealing with financial or personnel records) to devices that have an internal IP address. Nomadic users can access such services by using a Virtual Private Network (VPN), as is illustrated in Fig. 5.4. The figure shows a corporate network (left), which is the home network in this case and a hotel network (right), which is the foreign network. A mobile device in the hotel network (blue circle) wants to connect to servers in corporate network (brown circles) but it needs an internal IP address. This is done by setting up a VPN between the device and the corporate network. The VPN can be thought of as a virtual link that connects the mobile device to the corporate network at the IP level, so it has an internal IP address.

Concretely, the VPN is realized using IP encapsulation as shown in Fig. 5.3. The IP addresses in the outer IP header are the associated with the VPN server in the corporate network and a virtual network interface on the mobile device (top yellow circle). When the mobile device establishes the VPN, it is assigned an IP address with the network prefix of

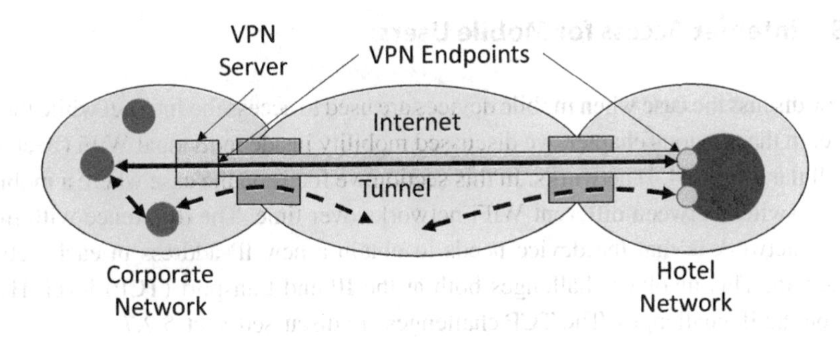


**Fig. 5.4** Virtual private network

the corporate network. This means that the two IP addresses in the inner IP header both have the prefix of the corporation, so the from server's perspective, the mobile devices is part of the corporate network. The VPN link also guarantees the confidentiality and integrity of the data it carries.

In most use cases, mobile devices need to use the VPN only for destinations inside the corporate network (black arrow). Traffic to public servers in the Internet use the regular network interface (bottom yellow circle) that bypasses the VPN (dashed arrow). An alternative is to send all traffic that is not local to the home network through the VPN. This type of "full" VPN may be useful when the mobile device wants to access a public server using a corporate IP address or when it wants to hide what servers it is accessing to observers in the hotel network. One drawback of using a full VPN is that it can create very long paths. For example, a mobile device in Europe is used to reserve a table in a local restaurant using a full VPN to a network in the US.

### Cellular

Cellular devices need a Subscriber Identity Module (SIM) to connect to a cellular network. The SIM identifies the account a user has with a cellular operator. Switching between cellular operators require replacing the SIM, which is a manual process. Fortunately, major cellular operators have very extensive coverage and excellent support for mobility in their coverage area, so there is rarely a need for switching between operators.

One exception is when traveling internationally, for example from the US to Europe. Some cellular operators have agreements with other operators that allow their customers to us their network. This is done automatically and it does not require changing SIMs. Alternatively, users can replace their SIM with the SIM of a local operator. This may be a cheaper solution compared with using an international plan through their "home" operator. In both cases, users will be disconnected for some period of time.

### 5.1.3 Internet Access for Mobile Users

We now discuss the case when mobile devices are used to access the Internet while the user is mobile. In the previous chapter, we discussed mobility inside individual WiFi (Sect. 4.3.11) and cellular (Sect. 4.4.4) networks. In this section we focus on the case where a mobile user needs to switch between different WiFi networks over time. The difference with mobility inside a network is that the device needs to obtain a new IP address in each network it connects to. This involves challenges both at the IP and transport (TCP) level. Here, we focus on the IP challenges. The TCP challenges are discussed Sect. 5.2.7.

The main challenge for a mobile WiFi device that is roaming between two WiFi networks it that it needs to connect to a new network, which is a multi-step process (Sect. 4.3). For example, it requires the device to associate with a new basestation, authenticate, and obtain information about the new network using DHCP. Each of these operations can be time consuming (Pei et al., 2017b). Practically speaking, this means that there will be a period during which the user is not connected to any network, resulting in a noticeable disruption in network connectivity.

Some research projects have designed techniques to support mobile WiFi users. One possibility is to connect to two WiFi networks at the same time, for example by using two WiFi cards or by using techniques such as MultiNet (Chandra et al., 2004). In that case, it may be possible to remain connected to the old WiFi network while connecting to the new WiFi network. It requires that the overlap in the coverage areas of the two WiFi networks is large enough so that the delay associated with connecting to the new network can be hidden. A number of research projects have studied how to use WiFi for vehicular applications. This may be possible when driving in areas that are entirely covered by a single WiFi provider, but switching between APs fast enough to maintain continuous connectivity remains a challenge (Deshpande et al., 2009; Hare et al., 2012).

**Dual-homed devices**—Finally, cellphones today typically have both a cellular and WiFi interface. As a result, they can connect to two types of networks, possibly at the same time. Such devices are called *dual-homed*, or more generally *multi-homed*, independent of whether the connections are wireless or wired. Dual-homed devices should generally be able to be connected to at least one network. Note that the two interfaces will have different IP addresses since they are associated with different networks (different Network IDs), which may impact TCP as discussed in the next section.

## 5.2 Impact of Wireless on Application Throughput

Communication session that include a wireless link can suffer from higher latency and packet loss rate, both of which can impact application-level throughput. In this section we provide background on two factors that determine network throughput: (1) how flows share resources (link bandwidth) in the network independent of what protocols are used on the endpoints (Sect. 5.2.1 ) and (2) how TCP congestion control selects an appropriate transmit rate for a flow. We then discuss how RTT and packet loss impact application throughput. Finally, we explain how TCP is impacted by IP mobility and introduce multipath TCP as an example of a solution.

### 5.2.1 Bandwidth Sharing in a Network

In this section we look at how flows in a network share bandwidth on links, ignoring the question of how senders can learn how much network bandwidth is available to them.

Broadly speaking, bandwidth allocation has two goals. The network wants to be efficient in the sense that it maximize the total aggregate bandwidth it serves to customers. A second goal is that bandwidth allocation should be "fair" so all customers have comparable performance. Let us first consider a simple network with a "dumbbell" topology as shown in Fig. 5.5. All links have the same bandwidth $B$ and $N$ flows share the link $L$ connecting the two routers R. $L$ is the *bottleneck link*, i.e., the link that limits the throughput of a flow, for all $N$ flows. In this case, a bandwidth allocation $T$ for each flow that is both efficient and fair is simply

$$T(A) = \frac{B}{N} \tag{5.1}$$

Figure 5.6b shows a richer topology with five flows numbered 1...5. All links are full-duplex with bandwidth $L$ in both directions. This topology is slightly more realistic since the various flows face very different network conditions. First, flows have paths with different numbers of hops. Moreover, links have different numbers of flows competing for the link bandwidth $L$. One possible bandwidth allocation is to assign all flows the same bandwidth, specifically $L/3$. This allows flows 1, 2, and 3 to evenly share and fully utilize the bandwidth

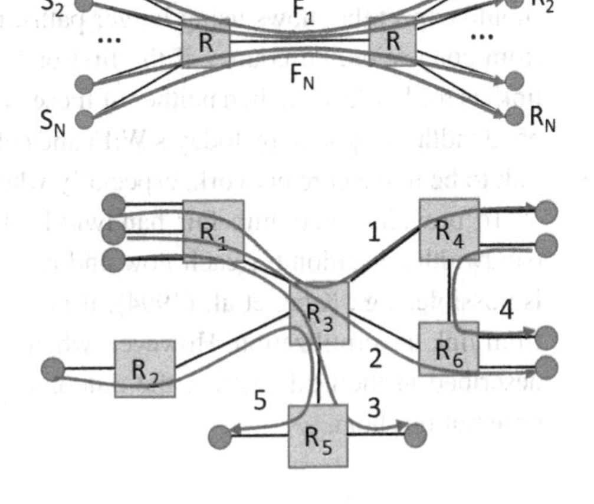


**Fig. 5.5** Competing flows in dumbbell topology

**Fig. 5.6** Bandwidth sharing in general topology

on link $R_1 \rightarrow R_3$, which is the most congested link. While fair, this allocation is not efficient since flows 4 and 5 could use more bandwidth without hurting the throughput of flows 1, 2 and 3. We need a bandwidth allocation algorithm that better balances fairness and efficiency.

A widely accepted fairness definition for networks is called Max-Min fairness (Le Boudec, 2005). Its goal is to maximize the throughput of flows that face the most challenging conditions, e.g., they use links with a lot of competing traffic (e.g., $R_1 \rightarrow R_3$), while allowing "lucky" flows to get a higher throughput. Informally, with a Max-Min fair bandwidth allocation, it is not possible to increase the throughput of any flow by reducing the bandwidth of a flow that has a higher throughput, so bandwidth is first allocated to flows that face the most challenging conditions.

Focusing on a simple network where all links have the same bandwidth, a max-min fair allocation can be calculated by first assigning bandwidth to flows that share the most congested link, i.e., the bottleneck link for flows that will have the lowest throughput. In our example, the three flows that share link $R_1 \rightarrow R_3$ get throughput $L/3$. We then allocate bandwidth for the next most congested link, while taking into account that some bandwidth has already been allocated. In our example, this is link $R_3 \rightarrow R_5$, where, considering the bandwidth already allocated to Flow 3, $2 \times L/3$ bandwidth remains for Flow 5. Finally, Flow 4 is assigned a bandwidth of $L$. More generally, Max-Min fair bandwidth is allocated iteratively using the following equation:

$$T(A) = \frac{B(L) - \sum_{i=1} OT_i}{N_h} \tag{5.2}$$

It is similar to Eq. 5.1, except that we need to reduce $B(L)$ with the throughput $T_i$ of the O flows that are bottlenecked elsewhere.

The results in Fig. 5.6 provides some interesting insights, besides the obvious observation that flows that use highly congested links will have lower throughput. First, how much throughput a flow gets depends not only on what links it uses, but also on the traffic in other parts of the network. For example, Flow 5's throughput depends on how many flows use link $R_1 \rightarrow R_3$. If Flow 1 terminates, Flow 5's throughput will decrease! Second, one should expect that flows using longer paths, i.e., with more hops, are more likely to suffer from congestion. Of course, if the first or last link on the path (which may be a wireless link) is the bottleneck, then neither of these observations apply. Note however that given the bandwidths supported by today's WiFi and cellular, it is certainly possible for the bottleneck link to be in the core network, especially when the wireless network has low utilization.

To provide a max-min fair bandwidth allocation the network must calculate the fair bandwidth allocation for each flow and provide this information to the sender. While this is possible, e.g., Kung et al. (1994), it is not practical on an Internet scale and can suffer from link underutilization. However, when senders adapt to congestion in the network, as described in the next section, the outcome is similar a max-min allocation so the above observations hold.

### 5.2.2 TCP Basics

Reliable byte streaming protocols such as TCP are by far the dominating transport protocol used by applications. TCP supports four functions: connections establishment and tear down, flow control, error control, and congestion control. TCP is a *connection-based protocol* so the two endpoints maintain state that is used to optimize bi-directional, reliable data transfers. Connections are established and torn down using an exchange of control packets. The goal of flow*flow control* is to ensure that the TCP sender never sends data unless the receiver has buffer space to store it. It is implemented by having the receiver pace the sender by advertising how much free buffer space it has. The *error control* function makes sure that the data transfer is reliable. This requires that the sender can detect and recover from lost and corrupted packets. TCP uses mechanisms such as acknowledgements and retransmission for error control. Finally, the *congestion control* function controls the transmit rate of the sender to avoid, or control, congestion on network links. In contrast to the previous three functions, its operation can be impacted by the presence of wireless links in the TCP network path. In the next section, we give an overview of how TCP congestion control works and we then discuss how packet loss and roundtrip time impact congestion control, and thus the throughput of a TCP flow.

QUIC (Langley et al., 2017) is a transport protocol developed and used by Google. It implements the same functions as TCP does, but it uses TLS (Rescorla, 2018) for security by default and also implements some features of HTTP2 (Belshe et al., 2015; Stenberg, 2014a). Both TCP and QUIC can be configured to use a wide range of congestion control algorithms.

### 5.2.3 TCP Congestion Control

TCP Congestion Control (CC) controls the transmit rate with the goal of maximizing application throughput without introducing significant congestion in the network. Short term congestion is unavoidable and it is managed by having queues in the routers that buffer packets when traffic exceeds the link capacity. However, long term congestion results in significant packet loss that hurts performance. It can even result in network collapse (Nagle, 1984). TCP uses a Congestion Control Algorithm (CCA) that calculates a maximum transmit rate for the TCP session based on implicit and sometimes explicit feedback (Floyd & Jacobson, 1993; Ramakrishnan et al., 2001) from the network. Congestion control has been an active area of research for over 30 years and many CCAs are in use today (Mishra et al., 2020), but they all have a similar structure. We focus on an older CCA, *TCP new Reno* (Stevens, 1997), since it is the basis for many of the CCAs in use today.

At the start of a new TCP session, or after a TCP session has been idle, the TCP sender has no information about the available bandwidth on the network path to the receiver, so it uses aggressive probing, called *Slow Start* (SS), to quickly get an estimate of the available bandwidth. During SS, the sender starts with a fixed initial transit rate, and then doubles the transmit rate every roundtrip time (RTT). When the traffic load exceeds the available bandwidth on the path, the router queue on the bottleneck link will fill up, which will eventually result in packet loss. The TCP sender interprets packet loss as a *congestion event*, an indication that one of the links on the path is congested, and it reduces its transmit rate to limit congestion.

After a packet loss, TCP switches to a *Congestion Avoidance* (CA) phase in which it tries to "track" the available bandwidth, since it changes over time as we discussed in the previous section. This is implemented by slightly increasing the transmit rate in each RTT, effectively probing for more bandwidth. These small rate increases by all TCP senders sharing a congested link will eventual result in queue build up on the bottleneck router, leading to packet loss. In response, the sender reduces its transmit rate and then immediately starts probing again for more bandwidth.

TCP does not control the transmit rate directly. Instead, it maintains a variable called *congestion window* ($cwnd$) which represents the amount of data it can transmit in one RTT. The transmit rate, represented by $cwnd/RTT$, is controlled by increasing or decreasing $cwnd$. In practice the RTT can change due to packet queueing in the routers, but $cwnd$ is often a good proxy for throughput. How the transmit rate is controlled may look like a minor implementation detail, but it in fact has significant impact on the throughput.

Figure 5.7 shows how $cwnd$ changes over time for TCP new Reno. During the SS phase, $cwnd$ doubles every RTT, so the rate increases exponentially. Most TCP versions today use an initial window of 10 MSS (Maximum Segment Size, typically 1448 Bytes). During the CA phase, $cwnd$ increased by one MSS every RTT, so the transmit rate increases linearly. Finally, after a packet loss, TCP cuts $cwnd$ in half. This *Additive-Increase, Multiplicative-Decrease algorithm* for adjusting the transmit rate during the CA phase has been shown to converge to a steady state in which senders share the bandwidth equally under certain conditions (Chiu & Jain, 1989).

Since TCP New Reno, many improvements have been made to congestion control. One major change is that some CCAs also interprets RTT increases as a congestion events, since it indicates queue build up (Brakmo et al., 1994). For example, Cubic (Ha et al., 2008; Rhee et al., 2018), the most widely used CCA in use today (Mishra et al., 2020), uses RTT increases as a congestion event for exiting SS (Ha & Rheeu, 2011). It also adjusts the transmit

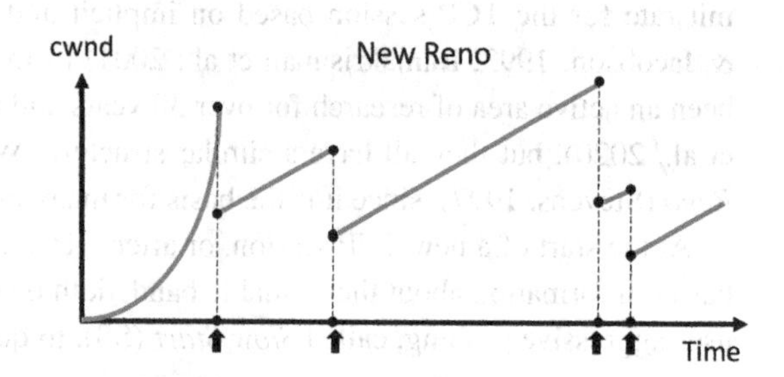


**Fig. 5.7** TCP congestion control window (cwnd) during a TCP session

rate differently during both the SS and CA phases. BBR (Cardwell et al., 2022, 2016), the second most common CCA, tries to limit queue build up, which contributes to the RTT, so it primarily relies on RTT increases as a congestion signal.

### 5.2.4 Impact of RTT on Throughput

How TCP increases and decreases its transmit rate has a significant impact on network throughput. The reason is that the throughput is a time-based metric ($bits/sec$) while TCP transmit rate changes are based on RTT ($bits/RTT$). That means that low-RTT flows increase their transmit rates more rapidly during both SS and CA compared to high-RTT flows.

To illustrate this, Fig. 5.8 shows the transmit rate of two flows during SS: Flow 2 (red) has an RTT that is twice that of Flow 1 (blue). The X axis shows time; the RTT of Flow 1, $RTT_1$, is used as the unit of time. The y-axis shows transmit rate (Mbps). We see that the throughput of Flow 2 is consistently lower than that of Flow 1 and that difference increases over time. There are three reasons for this. First, Flow 2 starts at a lower transmit rate than Flow 1 since TCP uses the same initial $cwnd$, irrespective of RTT. For example, for an $RTT_2 = 50$ ms, Flow 1 will start to transmit with a rate of 0.23 Mbps, while Flow 2 will start at 0.115 Mbps. Second, Flow 1 doubles its transmit rate twice as often as Flow 2. As a result, after 201 ms, Flow 1 will transmit at a rate of 3.7 mbps ($cwnd$ has doubled 4 times), while Flow 2 will only have a rate of 0.46 Mbps ($cwnd$ has doubled only twice). Finally, assuming the two flows use paths with a similar available bandwidth, Flow 2 will stay in Slow Start much longer than Flow 1, so it takes Flow 2 longer to reach the CA phase, during which transmit rates are higher.

Flows with a lower RTT similarly have an advantage during the CA phase. An analysis of TCP New Reno (Mathis et al., 1997) for a simple dumbbell topology (see Fig. 5.5) shows that the throughput of a TCP flow during the CA phase can be represented as

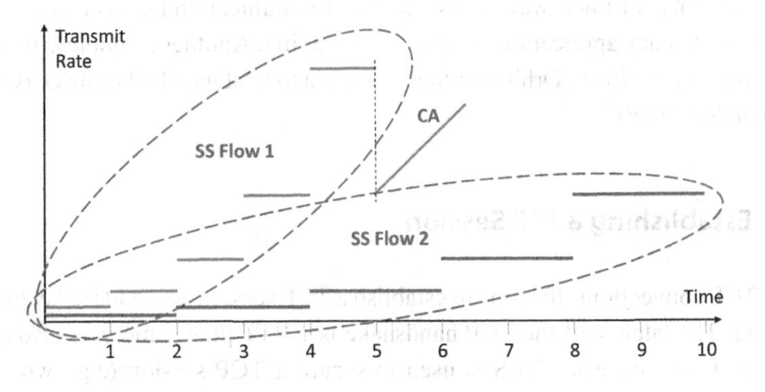


**Fig. 5.8** Impact of RTT on TCP transmit rate during slow start

$$T(A) = \frac{MSS}{RTT} \times \sqrt{\frac{8}{3 \times p}} \tag{5.3}$$

where $p$ is the packet loss rates, or more generally, the rate of congestion events. In steady state, the transmit rate is inverse proportional to the RTT, assuming the same packet loss rate, so again low-RTT flows have an advantage over high-RTT flows. The reason is that low-RTT flows are more aggressive than high-RTT flows since they increase *cwnd* faster. Follow on research generalized this result, and the impact of both the RTT and loss rate remains the same (Padhye et al., 1998; Philip et al., 2021).

While today's TCP versions use much more sophisticated CCAs than TCP Reno, the above results still largely hold. First, the SS phase has not changed much. It increases the transmit rate very quickly at a pace that is measured in RTT units, so low RTT flows continue to have an advantage. Similar, almost all CCAs regularly probe for more bandwidth using a clock that is RTT-driven. One exception is the BBR, which uses a very different algorithm to control the transmit rate (Cardwell et al., 2016).

WiFi and cellular access networks generally have higher latency than wired access links. For example, for WiFi, (Sui et al., 2016) the use of carrier sense to gain access to the shared transmission medium adds delay, especially on busy networks. Power save mode (Sect. 4.3.8) can also add significant delay to AP-to-device packet transmissions. Cellular networks have historically also had very long delays. Two reasons are that the control plane for cellular is complex so signaling is slow. In addition, cellular devices need to request bandwidth resources (Sect. 4.4.6) from the basestation before they can transmit, which further increases latency (Tan et al., 2021; Huang et al., 2013; Zhang et al., 2019). Unfortunately, this means that for most CCAs used in TCP today, network throughput will be lower compared with senders using wired access links. These same factors can also make latency, and throughput, more variable on wireless links.

Interesting enough, replacing wired by wireless links can in some cases lead to lower delays. The reason is that the speed of electro-magnetic signals is about 30% slower in fiber than in free space. This can make a big difference for long distance communication. One example is the use of microwave networks for communication between cities (Rhee et al., 2022) for low latency applications such as stock trading. Another example is the deployment of a network of Low Earth Orbit satellites to form a low latency ISP in space (Giuliari et al., 2020; Handley, 2018).

### 5.2.5 Establishing a TCP Session

For new TCP connections, the time to establish a TCP session also adds delay to the delivery of the data. The latency of the TCP handshake is 1 RTT plus some overhead on the client and server. However, when TLS is used to secure a TCP session (e.g., when HTTPS is used to browse the web) an addition latency of two (TLS 1.2) or one RTT (TLS 1.2) is incurred, in addition to the cost of the cryptographic operations that must be performed on both endpoints.

Another consideration is that many applications establish a TCP session based on a DNS name, not an IP address. In that case, DNS name resolution, i.e., translating the domain name into an IP address, results in an additional delay, which is highly variable. In the best case, the name-address mapping is cached on a local DNS resolver in the client's network. In the worst case, the name needs to be retrieved from the DNS server of intended TCP destination, which may involve accessing several DNS servers throughout the Internet.

### 5.2.6 Impact of Packet Loss

Most versions of TCP interpret packet loss as a congestion event, on the assumption that packet loss is the result from congestion-induced queue overflow. Early versions of WiFi had high packet loss rate. Unfortunately, this resulted in very low throughput for TCP flows for two reasons. First, packet losses were detected by timeouts, which stalled packet transmission for as much as a second. Second, packet losses were interpreted by TCP as congestion events, so they cut *cwnd* in half. Researchers developed several fairly complicated techniques to hide, or efficiently recover from, wireless packet losses (Balakrishnan et al., 1996).

The solution that was ultimately adopted was in fact much simpler. Rather than fixing the problem, it is much easier to avoid the problem by designing wireless protocols with low packet loss rates, comparable to those on wired links. This can be done easily using techniques such as link-level packet retransmission and forward error correction (Chap. 4). This simple solution that is also faster and often more efficient. Link-level recovery is very fast since RTTs are very short. In contrast, end-to-end recovery at the transport level is much slower, since RTTs can be as high as 100s of ms.

### 5.2.7 Maintaining a TCP Session While Mobile

In Sect. 5.1.3 we discussed how inter-domain mobility results in a change in the IP address of the mobile device. Surprisingly, this has an impact on active TCP sessions. A TCP session is uniquely identified by a 5-tuple: the source IP address and port number, the destination IP address and port number, and the protocol number (TCP in this case). The address/port number pairs unique identify the sockets used by the communicating applications. This means that when one of the IP addresses changes, the identity of the TCP session changes. This confuses the end points since they use the 5-tuple to associate incoming packets with a specific TCP connection.

This problem is not fundamental. It is simply the result of an early design decision (Information Sciences Institute, 1981) that did not consider mobility. The problem can be solved by adding a layer of indirection (Snoeren & Balakrishna, 2000). TCP sessions are identified based on the IP addresses that the endpoints used when the session was established. When an IP address changes during the lifetime of the session, TCP can add an entry to a table that maps the 5-tuple of each incoming TCP packet to the 5-tuple used to establish the connection. The reverse mapping is performed for outgoing packets. This technique is however not widely support since it is only needed for WiFi (Sect. 5.1). The QUIC transport protocol, is an example of a transport protocol that supports mobility by design. During session establishment, QUIC creates a session identifier that is used to associate packets with QUIC sessions so changes in IP address do not affect its operation. Note that changes in the IP addresses require authentication to avoid hijacking of active TCP sessions.

### 5.2.8 Multi-path TCP

Multipath TCP (MPTCP) allows a single TCP session to make use of multiple Internet paths between the two communicating devices. Figure 5.9a shows a simple example. Nodes pairs A-B and C-D communicate over MPTCP connections, using two network interfaces on devices A and D, respectively. Figure 5.9b shows the high-level design of MPTCP (Ford et al., 2020; Murray et al., 2011). Similar to single-path TCP, MPTCP provides a reliable byte stream service between two applications through a socket. It is implemented as a module that distributes the data to be sent over one or more parallel single-path TCP sessions. The MPTCP module on the receiver places the data received over the parallel TCP connections in the correct order before making it available to the receiving application. Similar to TCP, MPTCP must implement four functions core TCP functions (Sect. 5.2.2). Using multiple paths can improve the throughput and reliability of a TCP session.

MPTCP sessions are established by first setting up a single path TCP session. The two endpoints also exchange control information needed to add and manage additional paths. At a later time, additional TCP connections can be established and the new connections can be added to an existing MPTCP connection using TCP protocol extensions. Individual TCP

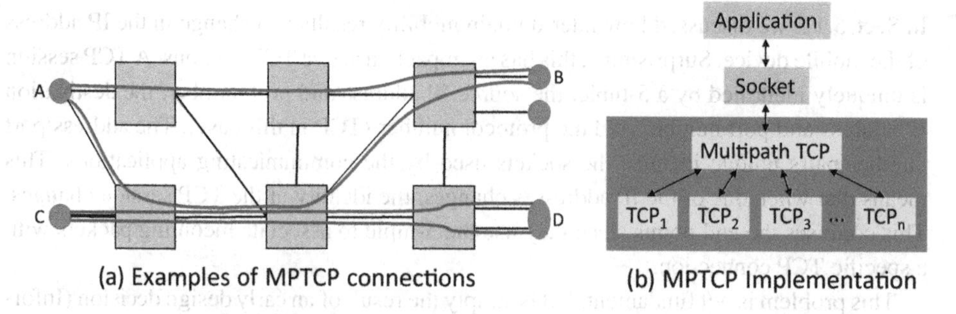


**Fig. 5.9** Multipath TCP

connections that are part of a MPTCP connection can be terminated at any time. Paths can be added and removed at the request the applications. As a result, the use of MPTCP is not transparent to applications. While communication uses a single socket, applications are responsible for adding and removing paths.

Each single-path TCP connection implements flow, error, and congestion control independently. This is important since the two paths typically have different properties, e.g., RTT, available bandwidth, etc. The MPTCP module on the sending node is responsible for distributing the data between the parallel TCP sessions. MPTCP should generally get a higher throughput than a single-path TCP connection (Chen et al., 2013). While this may be attractive to sending applications, one can argue that this is not fair. As a result, researchers have explored MPTCP-specific congestion control algorithms that can be used by the MPTCP module, for example, to achieve a throughput that is similar to that of a single path connection, e.g., Wischik et al. (2011).

MPTCP can be useful for cellular users and connected vehicles, since they often have both a WiFi and cellular network interfaces. Using a two-path MPTCP connection can help maintain network connectivity during WiFi handoffs or other path failures. When a path fails, the other path keeps the TCP session alive (Khan et al., 2022; Lim et al., 2014; Paasch et al., 2012). Some MPTCP implementations also allow application to control MPTCP features, e.g., by choosing a congestion control algorithm or by controlling how much or what data is transferred over each path based on its properties (Lee et al., 2018; Thakur & Kunte, 2020; Vu & Walker, 2019).

# 6 Applications in Wireless and Mobile Networks

## 6.1 Application-Centric Metrics for Wireless and Mobile Devices

We elaborate on the wireless network and device properties considered in this section:

- *Availability of network service*: Loss of network connectivity impacts all applications that use the network. For most applications, loss of network connectivity is fatal, although some applications can be used in disconnected mode, as we discuss in Sect. 6.2.
- *ADU delivery time*: Most applications do not operate on packets or byte streams. Instead, they generate or process Application Data Units (ADU), blocks of data such as a document, a sequence of video frames, or an OS upgrade. As a result, the time to deliver an ADU is an important metric. This metric depends on both the available bandwidth and RTT of the network path that is used to deliver the data, and both of these parameters are impacted by a wireless access link, as discussed in Sect. 5.2.
- *Mobile device resource constraints*: Mobile device often have limited processing resources, so applications may have slow response times or may not be able run at all. Many research groups have explored offloading of compute tasks, but this requires good network performance.

In this chapter, we discuss the impact of each of these above metrics on mobile applications and we also discuss techniques that can be used to mitigate them. The presentation is based on sample applications and is not meant to be exhaustive.



P. Steenkiste, *Introduction to Wireless Networking and Its Impact on Applications*,
Synthesis Lectures on Mobile & Pervasive Computing,
https://doi.org/10.1007/978-3-031-27466-4_6

## 6.2 Availability

We discuss wireless connectivity for both mobile and nomadic users and its impact on applications.

### 6.2.1 Wireless Coverage

Internet access based on a wired access link is very reliable. While outages do happen, for example due to fallen trees or equipment failures, they are rare. In contrast, disconnections are more common with wireless access links since access depends on devices being inside the coverage area of a basestation for which they have credentials. As a result, devices may not have internet access in certain areas, resulting in *intermittent connectivity* for the users. The likelihood and nature of disconnections depends on the access technology used and the type of mobility.

Near-ubiquitous coverage is an important goal when deploying cellular networks. The reason is that providers put a lot of emphasis on the Quality of Experience (QoE) of its customers: since users pay for cellular service, they expect high quality voice and good data throughput everywhere. Of course, no cellular provider can provide coverage everywhere. In areas that are very remote or inaccessible, or areas with very low population density, providing cellular coverage may be impossible or too expensive. Cellular providers typically document their coverage, e.g., on the web, so users know what service they can expect.

WiFi is also widely available but the infrastructure is owned by many organizations and individuals, e.g., homes, corporations, hotels, hotspots, etc. As a result, WiFi access is much more limited than cellular since users can only use WiFi networks for which they have the right credentials. As a result, WiFi coverage for a user is typically only available in specific locations.

### 6.2.2 Impact on Mobile Users

Mobile devices, especially cellphones, are often used while the user is mobile, e.g., making a call while walking or using a navigation applications while driving. Even small coverage gaps can result in intermittent connectivity that significantly impact user experience.

Cellular is designed to support mobile users so continuous connectivity is not a problem in most areas, even in challenging environments. As discussed in Sect. 4.4.8, each cellular generation sets a mobility goal corresponding to the fastest travel speed envisioned during the lifetime of that generation. A number of studies have evaluated cellular data access on high speed trains (Li et al., 2017; Pan et al., 2022; Wang et al., 2019), which can travel at

300 km/h and higher. The results of the studies show that connectivity is generally good, but at the highest speeds there are sometimes short 1 second gaps in connectivity that are the result of a failed handover. Unfortunately, these gaps can confuse the TCP protocol resulting in a significant drop in throughput. Other work (Li et al., 2018) shows that multipath TCP can help cover up the connectivity gaps since there is always one path that has good connectivity.

Even in areas with good WiFi coverage, such as a large campus with a managed WiFi network, connectivity may not be continuous. While there are WiFi standards that optimize connectivity for mobile users (Sect. 4.3.11), they are not always supported since the focus of deployments is typically nomadic users.

### 6.2.3 Disconnection Operation

Users of mobile devices can be disconnected from the network for extended periods, e.g., while traveling by air, foreign travel without WiFi or cellular access, or while walking with a laptop in a backpack. Clearly, interactive applications, such as voice/video conferencing, or applications that require access to data in the cloud, cannot be used in such cases. However, a number of applications rely on Internet access, but thay can still be used when disconnected. The two key ideas that enable disconnected operation are (1) caching relevant data on the mobile devices, and (2) propagating any updates to data to the server when the mobile device reconnects to the network.

#### Shared File Systems

The Coda File System (Kistler & Satyanarayanan, 1992) supports disconnected users who want to use or modify files in a shared file system such as AFS (Satyanarayanan, 1990). The first step is that users specify what part of the file system they may want to work on while disconnected, so it can be replicated (cached) on the mobile device. Once cached, users can use and modify the cached copy of the files using their normal file names. Since there are now multiple copies of the files, Coda needs a replication strategy that defines how cached files can be used. Coda uses an *optimistic* strategy which allows devices to make changes to files locally (Davidson et al., 1985). In contrast, a *conservative* strategy allows only one replica to modify a file, which is not practical in a disconnected scenario since it severely limits the availability of files.

Whenever a device is connected to the network, Coda propagates changes to cached files to the file system. However, when disconnected, changes to files are applied locally only, and they are propagated to the server when the device reconnects. Since Coda uses a optimistic replication strategy, there may have been conflicting updates by other users that need to be resolved. Coda only worries about conflicts relevant to the Unix file system model, specifically the case when both the cached and the file server copy of a file have been updated. These conflicts are detected and flagged for manual repair. Coda also provides a mechanism for introducing application-specific plugin code for automating the repair of such conflicts. Online file sharing services such as Box and Google Drive can be viewed as descendants from Coda. They typically provide very limited, if any, support for disconnected operation.

### Structured Data

Coda supported disconnected use of data where the unit of data is a file. The goal of disconnected operation was to sure that the file contents (the byte stream) is the same as if the user had been working online. However, many applications and services work with structured data or "objects", typed data structures that have a specific format and for which a limited number of operations are defined. A simple example is e-mail. E-mail messages have a specific format, and they can be created, sent, read, forwarded, etc. Note that once sent, an e-mail cannot be modified, so it is a write-once data structures. E-mail is stored on a mail server and accessed by users using an e-mail client. While users can of course not send or check e-mail while disconnected, many other operations are supported. For example, users can compose e-mail messages and queue them, so they can be transmitted when the device is connected to the Internet. They can also read e-mail that is cached on their mobile devices. Finally, they can manage e-mail, e.g, delete e-mail or move it between folders. Offline operations can be replayed on the server when the user is connected to the Internet.

The offline solution used for e-mail can be used by many other applications, for example a calendar or task list. It has data structures representing appointments and tasks, each of which can be created, read, deleted, cached, etc. In contrast to e-mail messages, these data structures can also be modified, but these operations can be recorded and replayed when the mobile devices connect to the Internet. Note that similar to a shared file system, multiple users can share a calendar and task list, raising the question of replica control. Applications generally use an optimistic strategy, for the same reasons as Coda. Given the limited set of operations that can be applied, most offline updates do not result in conflicts since they modify different objects, e.g., an appointment or task. Updates to the same object can result in conflicts, for example when one user moves a meeting while another user deletes it.

### Server-initiated operations

The discussion so far focused on operations that are initiated by the client. However, services may also want to contact clients. For example, a mail server may push new e-mail messages proactively to clients and similarly calendar services can push new or modified appointment to clients. Mobile WiFi clients with intermittent connectivity will often change their IP address, raising the question of how servers can contact the device to push information. Mobile IP Sect. 5.1.2 was designed to solve this problem but it is not used in practice. A simpler solution is to have the client application contact the server whenever the mobile device reconnects to the network.

## 6.3 Time to Send an Application Data Unit

### 6.3.1 ADU Delivery Time

For many applications, the time to deliver an Application Data Units (ADUs) is an important performance metric. It referred to as the Flow Completion Time, and on the surface, the FCT is simply

$$FCT = \frac{S_A}{B} \tag{6.1}$$

where $S_A$ is the size of the ADU (in bits) and $B$ is the network bandwidth that is available for the application inside the network (in bits/sec). As we discussed in Sect. 5.2.2, wireless devices often have lower throughput than devices using a wired access link. The reasons are the wireless access links use multi-access protocols, so users share the link capacity, while wired access is based on dedicated links. In addition, wireless access links have in higher delays, resulting in longer RTTs that slow down TCP (Sect. 5.2.3). However, several factors, other than available bandwidth and RTT, impact the ADU transfer time, as we discuss next.

The transmit rate used to send an ADU depends on the size of the ADU. For large ADUs, e.g., downloading a 1 GByte OS patch, almost all data will be sent while TCP is in Congestion Avoidance mode, so the FCT can be calculated using Eq. 6.1 with B defined as the average throughput in CA mode. However, many applications have short ADUs, for example interactive applications, such as Web browsing or computer games. For small transfers, all or most of the data is sent in TCP Slow Start mode, when throughput is lower than in Congestion Avoidance mode. In addition, the TCP session establishment overhead has a significant impact on short ADUs, unless ADUs can reuse a previously established session (Sect. 5.2.5).

### 6.3.2 Optimizing Content Delivery

Many applications are latency sensitive and are used on many devices, ranging from PCs and even TVs with large, high resolution displays to mobile devices with small screen sizes. This means that they need to perform well for a wide range of available bandwidths. In this section, we will focus on two applications, web browsing and video streaming, since they are widely used on mobile devices. We start with web browsing for which the main performance metric is the Page Load Time (PLT), the time it takes to display the full web page after the user clicks on a URL. We present four broad classes of techniques that can be used to optimize network performance, some of which apply to wired and wireless networks, while others are specific to wireless.

**Content Delivery Networks**

The first class of techniques is to move the servers closer to the clients, thus reducing the RTT. This is done by deploying Content Delivery Networks (CDNs), networks of smaller clouds that replicate content closer to clients (Dilley et al., 2002; Nygren et al., 2010). Large content providers such as Amazon or Microsoft operate their own CDNs, while other content providers can work with commercial CDN providers such as Akamai and Cloudflare. Some CDNs operate a relatively small number of CDN nodes that cover large areas (e.g., the North-East region in the US), while others operate CDN nodes in many cities, further reducing RTT.

CDNs face the challenge of how they can direct user requests to the "best" CDN, i.e., the one that will result in the best performance. This is done using a technique called *DNS redirect*. When a client does a domain name look up for a web server, e.g., *www.cool-content.com*, the DNS server of *Cool Content* uses the IP address of the device issuing the DNS request to identify the approximate geographic in which the device is located, for example using a database. This information is then used to pick the CDN instance closest to the client. This binding can be static, or can can be dynamic considering other factors such as network performance information collected by the CDN provider. For content providers that contract with a CDN, there are a variety of techniques that can be used to direct the DNS request to the CDN, e.g., URL rewriting or delegating DNS resolution (Schomp et al., 2020).

**Optimizing HTTP**

Today's web pages have many embedded objects (video, ads, etc.) and the web page layout is optimized for the user's display. As a result, downloading a web page involves not a single, large ADU, but many smaller ADUs, some of which are stored on different servers. Web browsing, and many other applications, use the HTTP protocol, which has been revised over time to adapt to the increasing complexity of web pages (Grigorik, 2013). The first widely used version, HTTP 1.1, used a separate TCP session for each object, which was sufficient since web pages were simple at that time. Unfortunately, as described in Sect. 5.2.5, this is inefficient for pages that have many short objects. HTTP 1.2 addresses this issue by allowing a single HTTP session to be used for multiple pipelined requests, so many web objects can be transferred over a single TCP session. However, this can result in *head-of-line blocking* (HOL), where a delay in handling one HTTP request on the server delays all later requests. Web browsers reduced the impact of HOL blocking by distributing HTTP requests over multiple parallel TCP sessions, which increases overhead. HTTP2 (Stenberg, 2014b) avoids the need for parallel TCP sessions by supporting parallel, prioritized streams of requests in a single HTTP session. This makes it possible to retrieve many objects from same server while avoiding HOL blocking over a single TCP session, which is very efficient.

### Adapting Content to the Device

Another way to optimize the PLT for web pages is to adapt the web page itself to the device and available bandwidth. This can be very effective since it can reduce the amount of data that needs to be sent to a mobile device. For example, many content providers have special web servers for serving mobile devices. These servers may serve web pages with fewer objects and with images with lower resolution, which is acceptable since mobile devices have smaller screens. The content provider can redirect requests from mobile devices to a server optimized for mobile devices based on information provided by the client at the start of an HTTP session (e.g., browser type) or based on the IP address of the client. Another example is that cellular providers sometimes reduce the size of images requested by mobile clients, which reduces both the load on the wireless access links and the FCT of the image.

### The Use of Proxies

Many companies provide mobile web sites that are optimized for cellular users. Others have deployed proxies that execute JavaScript and push web objects to mobile clients to reduce the page load time. However, a few projects have gone even further and developed a special web infrastructure for mobile web browsing.

One example is Google's Accelerated Mobile Project (AMP) which speeds up the page load time for the mobile web by restricting the format of the web page by using a simpler markup language. When a user requests an AMP-enabled web page using an AMP-enabled browser, the page can be very quickly rendered and displayed. The rendered pages are cached and served from Google's CDN, further improving performance. AMP leads to lower page load times and time to first byte for AMP-enabled web pages (Jun et al., 2019), but at the cost of simpler pages.

Another example is Nutshell (Sivakumar et al., 2017). Using proxies can improve the user experience but it can also places a high burden on the proxies, potentially making them a bottleneck. Nutshell introduces a technique called *whittling*, which partitions the execution of Javascript between the mobile device and the proxy. The proxy executes only the JavaScript code that is needed to identify embedded objects so it can fetch and push these objects to the mobile device. The mobile device executes the full Javascript so the viewed web page is not affected by the optimization. Nutshell reduces the load on the proxy while preserving the latency benefits of a traditional web proxy.

### Video Delivery

For video streaming, the primary metrics are avoiding video stalls (the video freezes), so video frames must be delivered to the client before their replay time, and optimizing the video resolution (video bit rate). Video streaming today is supporting using the HTTP protocol. Specifically, video is sent as a sequence video "segments" that contain a group of video frames. One of the reasons why video is served this way is that it allows continuous adaptation of the video bit rate based on the available bandwidth. The key idea is that the client continuously estimates the available bandwidth based on the download time of recently downloaded segments. It can then use these estimates to select the highest bit rate that can be supported without running the risk of stalling the video.

Specifically, when a user clicks on a video link, a *video manifest* file is downloaded. It lists all the information needed to download and play the video, including for each segment, a list of URLs for files that encode the video segment with different resolutions. The file size increases with the resolution. The player starts by downloading the first few segments of the video, typically using a low bit rate since it does not yet know the available bandwidth. Once it has enough chunks to avoid video stalls, it starts to play the movie. Segments have a certain duration $D$ so the download time of segments should comfortably fit in $D$. The player can estimate the available bandwidth, for example based on sliding window average of the throughput observed for earlier segments, to pick the resolution for the next segment. The algorithm generally also considers other factors to optimize user QoE, for example, it avoids switching resolution too frequently (Jiang et al., 2012). Video streaming providers often also rely on *video brokers* that select for each segment the best CDN instance from potentially several CDN providers, based on network performance, as observed by video players in the same part of the Internet (Mukerjee et al., 2017).

While the above technique is specific to video, it can be used by any application that periodically sends or receives ADUs over a longer period of time.

## 6.4 Resource Constrained Mobile Devices

Mobile devices are more resources constrained than stationary devices such as desktop computers and servers. While today's laptops are very capable, cells phones, wearable devices, and embedded devices generally have more limited CPU capabilities. While it is possible to build mobile devices with more powerful CPUs, this is often not practically because of cost, weight, and power constraints.

### Computational Offloading

Computational offloading has long been recognized as an attractive way of supporting applications on resource-constrained devices with a high quality of experience for the user. An early paper (Noble et al., 1997) not only motivates the need for offloading but it also argues that it should be *adaptive* since the availability of both network and compute resources in the infrastructure is variable. The paper proposes Odyssey, a system that supports adaptation to resource variability. Since then, computational offloading, sometimes also called *cyber foraging*, has continued to be a very active area of research (Flinn, 2012).

### Edge Computing

Today, offloading computation to the cloud, which offers an elastic pool of inexpensive compute cycles, is a common solution. However, traditional clouds can be very far from clients and as discussed in Sect. 5.2.4, longer paths and higher RTTs often result in lower network performance. A 2009 paper (Satyanarayanan et al., 2009) makes the case that mobile computing should not be limited to having ubiquitous access to information, but that should also help users use compute-intensive applications such speech recognition and augmented reality that are latency sensitive. They identify resource-poor mobile devices and the delay of communicating with remote clouds as major obstacles to realizing this vision. The proposed solution is to offload computation to *cloudlets*, computing clusters with good network connectivity serving nearby devices. The paper also proposes to use VMs to create customize environments for client devices to simplify offloading. Follow on work shows the importance of using cloudlets instead of remote clouds by quantifying the impact of network RTT on application latency and user satisfaction for a set of diverse application (Chen et al. (2017) and Satyanarayanan (2017)). Today, edge computing is a growing area both in industry and academia.

In remainder of this section, we discuss offloading for two classes of devices, cellphones and smart vehicles, as examples of embedded device that must support a variety of applications. We will use the term "cloud offloading" for simplicity. The cloud can be a remote cloud or a nearby cloudlet. although for most applications, cloudlets will be needed to achieve good performance.

### Cellphones

A number of research projects have proposed solutions for offloading computation from cellphones to the cloud. One example is Odessa (Ra et al., 2011), which speeds up interactive perception applications through parallel execution on a cellphone and a server. Odessa uses a data flow graph representation of the application to identify opportunities for offloading, and for data and pipeline parallelism. Offloading decisions are dynamic so the system can adapt to the scene complexity as well as network and device performance. The evaluation shows that the performance improvement can be as high as a factor of 3. Clonecloud (Chun et al., 2011) similarly exploits both data and pipeline parallelism to speed up applications and it uses static analysis and dynamic profiling to automatically optimize the application. Tango (Gordon et al., 2015) executes the application on both the device and a server. It returns the result of the execution that finishes first. While this brute force approach is resource intensive, it avoids the challenging tasks of partitioning the application and it always has the best response time.

Cellphones not only have limited compute cycles, but they are also energy constrained. Maui (Cuervo et al., 2010) offloads parts of the computation with the goal of reducing the energy consumption on the cellphone. The optimization problem that decides on what to offload is formulated as an ILP that takes into account both the energy for computation and communication. The authors also found that Power Save Mode (PSM) (Sect. 4.3.8), which is design to save energy, not only adds latency but can also increase energy consumption.

**Vehicular Computing**

Today's vehicles rely heavily on applications ranging from enhancing driver experience (e.g., voice recognition), driver assistance (e.g., blind spot detection), and increasingly safety-critical applications (e.g., platooning, autonomous driving). The applications generally use an on-board compute infrastructure, which has the advantage that applications are easy to develop and have predictable response times. Unfortunately, as vehicles become smarter, more on-board compute cycles are needed, increasing the cost of the vehicle. Even worse, vehicles have a lifetime of 10–15 years—clearly an on-board compute infrastructure will be outdated and inadequate much sooner. Offloading to the cloud offers and attractive alternative. However, in contrast to cellphones, vehicles often need to run several applications in parallel. In addition, applications sometimes consume a lot of data, e.g., video or lidar, so offloading the entire application is not practical. We now consider two examples of partitioning the application between the vehicle and the cloud.

A first system optimizes the execution of a set of driver-assist applications such as gesture recognition or identifying traffic signs (Ashok et al., 2017). The system receives a continuous stream of requests, each with a deadline. The key insight driving the design of the scheduler is that many interactive, sensor-driven applications are structured as a pipeline of stages and the size of the data that is passed between them shrinks with every stage. This means that we can adapt to the available network bandwidth by changing how many stages are executed on the vehicle, i.e., we can trade on-vehicle compute cycles for network bandwidth. Using this tradeoff, the scheduler calculates a partition for each active request using a policy that either minimizes network bandwidth or on-vehicle computer cycles.

Many offloading frameworks target applications that are stateless, in the sense that they do not maintained state across tasks. Stateful applications that use and update a significant amount of state are much harder to partition, since maintaining a consistent view of the state between the vehicle and the cloud generates a lot of traffic. This is especially an issue for latency sensitive applications. An example is Simultaneous Mapping and Localization (SLAM), an application that uses stereo vision or lidar input to continuously calculate the vehicle's position and orientation, called the pose. It maintains a complex data structure that is used track changes in the environment as the vehicle moves. SLAM also uses data from previous trips to periodically optimize accuracy.

CloudSLAM (Wright et al., 2020) partitions the OrbSLAM application by replicating the state. The vehicle executes *tracking*, which incrementally calculates a new pose for each frame. It also updates its copy of the data structure, using only data collected during this trip. Tracking is latency sensitive and lightweight. Unfortunately, tracking by itself results in drift, which accumulates over time. To avoid this, the vehicle regularly sends a frame to the cloud, which optimizes the pose based on its copy of the data structure, which includes previous trip data. This is an expensive operation. The cloud then sends the optimized pose to the vehicle, which uses it to update its copy of the data structure. CloudSLAM is adaptive in two ways. First, it sends a frame to the cloud based on distance traveled, not time. The reason is that distance better reflects the degree of change in the frame compared with time. This reduce the load on the network. Second, the vehicle compares the optimized pose received from the cloud with its own value of the pose. The difference represents the error associated with local pose estimation. When this error is above a threshold set by the user, CloudSLAM increases the rate at which frames are sent to the cloud, and when the error is low, it decreases the rate.

# Bibliography


Aboba, B., & Calhoun, P. (2003). RADIUS (Remote Authentication Dial In User Service) support for Extensible Authentication Protocol (EAP) . IETF RFC 3579. https://tools.ietf.org/html/rfc3579.

Abramson, N. (1970). THE ALOHA SYSTEM: Another alternative for computer communications. In *Proceedings of the November 17–19, 1970, Fall Joint Computer Conference (AFIPS '70 (Fall))* (pp. 281–285). New York, NY, USA: Association for Computing Machinery. https://doi.org/10.1145/1478462.1478502.

Aggarwal, S., Thirumurugan, A., & Koutsonikolas, D. (2019). A first look at 802.11ad performance on a smartphone. In *Proceedings of the 3rd ACM Workshop on Millimeter-Wave Networks and Sensing Systems (mmNets'19)* (pp. 13–18). New York, NY, USA: Association for Computing Machinery.

Akella, A., Judd, G., Seshan, S., & Steenkiste, P. (2005). Self-management in chaotic wireless deployments. In *Proceedings of the 11th Annual International Conference on Mobile Computing and Networking (MobiCom '05)* (pp. 185–199). New York, NY, USA: Association for Computing Machinery. https://doi.org/10.1145/1080829.1080849.

Ashok, A., Steenkiste, P., & Bai, F. (2017). Vehicular cloud computing through dynamic computation offloading. *Computer and Communications,120*(C), 125–137.

Balakrishnan, H., Padmanabhan, V. N., Seshan, S., & Katz, R. H. (1996). A comparison of mechanisms for improving TCP performance over wireless links. In *Conference Proceedings on Applications, Technologies, Architectures, and Protocols for Computer Communications (SIGCOMM '96)* (pp. 256-269). New York, NY, USA: Association for Computing Machinery. https://doi.org/10.1145/248156.248179.

Beard, C., & Stallings, W. (2016). *Wireless communications networks and systems*. London, UK: Pearson.

Belshe, M., Peon, R., & Thomson, M. (2015). Hypertext transfer protocol version 2 (HTTP/2). IETF RFC 7540. https://datatracker.ietf.org/doc/html/rfc7540.

Brakmo, L. S., O'Malley, S. W., & Peterson, L. L. (1994). TCP vegas: New techniques for congestion detection and avoidance. In *Proceedings of the Conference on Communications Architectures, Protocols and Applications (SIGCOMM '94)* (pp. 24–35). New York, NY, USA: Association for Computing Machinery. https://doi.org/10.1145/190314.190317.







P. Steenkiste, *Introduction to Wireless Networking and Its Impact on Applications*, Synthesis Lectures on Mobile & Pervasive Computing, https://doi.org/10.1007/978-3-031-27466-4

Bychkovsky, V., Hull, B., Miu, A., Balakrishnan, H., & Madden, S. (2006). A measurement study of vehicular internet access using in situ Wi-Fi networks. In *Proceedings of the 12th Annual International Conference on Mobile Computing and Networking (MobiCom '06)* (pp. 50–61). New York, NY, USA: Association for Computing Machinery. https://doi.org/10.1145/1161089.1161097.

Camp, J., & Knightly, E. (2008). Modulation rate adaptation in urban and vehicular environments: Cross-layer implementation and experimental evaluation. In *Proceedings of the 14th ACM International Conference on Mobile Computing and Networking (MobiCom '08)* (pp. 315–326). New York, NY, USA: Association for Computing Machinery. https://doi.org/10.1145/1409944.1409981.

Cardwell, N., Cheng, Y., Yeganeh, S. H., & Jacobson, V. (2022). BBR congestion control. IETF experimental draft. https://datatracker.ietf.org/doc/html/draft-cardwell-iccrg-bbr-congestion-control.

Cardwell, N., Cheng, Y., Gunn, C. S., Yeganeh, S. H., & Jacobson, V. (2016). BBR: Congestion-based congestion control: Measuring bottleneck bandwidth and round-trip propagation time. *Queue, 14*(5), 20–53. https://doi.org/10.1145/3012426.3022184.

Cerf, V., & Kahn, R. (1974). A protocol for packet network intercommunication. *IEEE Transactions on Communications, 22*(5), 637–648.

Chandra, R., Bahl, P., & Bahl, P. (2004). MultiNet: Connecting to multiple IEEE 802.11 networks using a single wireless card. In *IEEE INFOCOM 2004* (pp. 882–893). Piscataway, NJ: IEEE. https://doi.org/10.1109/INFCOM.2004.1356976.

Chen, Z., Hu, W., Wang, J., Zhao, S., Amos, B., Wu, G., Ha, K., Elgazzar, K., Pillai, P., Klatzky, R., Siewiorek, D., & Satyanarayanan, M. (2017). An empirical study of latency in an emerging class of edge computing applications for wearable cognitive assistance. In *Proceedings of the Second ACM/IEEE Symposium on Edge Computing (SEC '17)* (Article 14, p. 14). New York, NY, USA: Association for Computing Machinery.

Chen, Y.-C., Lim, Y.-S., Gibbens, R. J., Nahum, E. M., Khalili, R., & Towsley, D. (2013). A measurement-based study of MultiPath TCP performance over wireless networks. In *Proceedings of the 2013 Conference on Internet Measurement Conference (IMC '13)* (pp. 455–468). New York, NY, USA: Association for Computing Machinery. https://doi.org/10.1145/2504730.2504751.

Chiu, D.-M., & Jain, R. (1989). Analysis of the increase and descrease algorithm for congestion avoidance in computer networks. *Computer Networks and ISDN Systems, 17*(1), 84–99.

Chun, B.-G., Ihm, S., Maniatis, P., Naik, M., & Patti, A. (2011). CloneCloud: Elastic execution Between mobile device and cloud. In *Proceedings of the Sixth Conference on Computer Systems (EuroSys '11)* (pp. 301–314). New York, NY, USA: Association for Computing Machinery.

CMU. (2020). Living Edge Lab. https://www.cmu.edu/scs/edgecomputing/index.html.

Cuervo, E., Balasubramanian, A., Cho, D.-K., Wolman, A., Saroiu, S., Chandra, R., & Bahl, P. (2010). MAUI: Making smartphones last longer with code offload. In *Proceedings of the 8th International Conference on Mobile Systems, Applications, and Services (MobiSys '10)* (pp. 49–62). New York, NY, USA: Association for Computing Machinery.

Davidson, S. B., Garcia-Molina, H., & Skeen, D. (1985). Consistency in a partitioned network: A survey. *ACM Computing Surveys, 17*(3), 341–370.

Deshpande, P., Kashyap, A., Sung, C., & Das, S. R. (2009). Predictive methods for improved vehicular WiFi access. In *Proceedings of the 7th International Conference on Mobile Systems, Applications, and Services (MobiSys '09)* (pp. 263–276). New York, NY, USA: Association for Computing Machinery. https://doi.org/10.1145/1555816.1555843.

Dilley, J., Maggs, B., Parikh, J., Prokop, H., Sitaraman, R., & Weihl, B. (2002). Globally distributed content delivery. *IEEE Internet Computing, 6*(5), 50–58. https://doi.org/10.1109/MIC.2002.1036038.

Droms, R. (1997). Dynamic host configuration protocol. IETF RFC 2131. https://datatracker.ietf.org/doc/html/rfc2131.

Flinn, J. (2012). *Cyber foraging - Bridging mobile and cloud computing*. Springer Link. https://link.springer.com/book/10.1007/978-3-031-02481-8.

Flores, A. B., Guerra, R. E., Knightly, E. W., Ecclesine, P., & Pandey, S. (2013). IEEE 802.11af: A standard for TV white space spectrum sharing. *IEEE Communications Magazine,51*(10), 92–100.

Floyd, S., & Jacobson, V. (1993). Random early detection gateways for congestion avoidance. *IEEE/ACM Transactions on Networking, 1*(4), 397–413. https://doi.org/10.1109/90.251892.

Ford, A., Raiciu, C., Handley, M., Bonaventure, O., & Paasch, C. (2020). TCP extensions for multipath operation with multiple addresses. IETF RFC 6824. https://datatracker.ietf.org/doc/html/rfc6824.

Foukas, X., Marina, M. K., & Kontovasilis, K. (2017). Orion: RAN slicing for a flexible and cost-effective multi-service mobile network architecture. In *Proceedings of the 23rd Annual International Conference on Mobile Computing and Networking (MobiCom '17)* (pp. 127–140). New York, NY, USA: Association for Computing Machinery. https://doi.org/10.1145/3117811.3117831.

Foukas, X., Nikaein, N., Kassem, M. M., Marina, M. K., & Kontovasilis, K. (2016). FlexRAN: A flexible and programmable platform for software-defined radio access networks. In *Proceedings of the 12th International on Conference on Emerging Networking EXperiments and Technologies (CoNEXT '16)* (pp. 427–441). New York, NY, USA: Association for Computing Machinery.

Garg, V. (2007). Wireless communications and networking. https://www.academia.edu/19503445/Wireless_Communications_and_Networking.

Giuliari, G., Klenze, T., Legner, M., Basin, D., Perrig, A., & Singla, A. (2020). Internet backbones in space. *SIGCOMM Computer Communication Review, 50*(1), 25–37.

Gordon, M. S., Hong, D. K., Chen, P. M., Flinn, J., Mahlke, S., & Mao, Z. M. (2015). Accelerating mobile applications through flip-flop replication. In *Proceedings of the 13th Annual International Conference on Mobile Systems, Applications, and Services (MobiSys '15)* (pp. 137–150). New York, NY, USA: Association for Computing Machinery.

Grigorik, I. (2013). Making the web faster with HTTP 2.0. *Communications of the ACM,56*(12), 42–49. https://doi.org/10.1145/2534706.2534721.

Halperin, D., Hu, W., Sheth, A., & Wetherall, D. (2010). 802.11 with multiple antennas for dummies. *SIGCOMM Computer Communication Review,40*(1), 19–25. https://doi.org/10.1145/1672308.1672313.

Hamming, R. W. (1950). Error detecting and correcting codes. *The Bell Systems Technical Journal, 29*(2), 147–160.

Handley, M. (2018). Delay is not an option: Low latency routing in space. In *Proceedings of the 17th ACM Workshop on Hot Topics in Networks (HotNets '18)* (pp. 85–91). New York, NY, USA: Association for Computing Machinery.

Hare, J., Hartung, L., & Banerjee, S. (2012). Beyond deployments and testbeds: Experiences with public usage on vehicular WiFi hotspots. In *Proceedings of the 10th International Conference on Mobile Systems, Applications, and Services (MobiSys '12)* (pp. 393–406). New York, NY, USA: Association for Computing Machinery. https://doi.org/10.1145/2307636.2307673.

Ha, S., & Rheeu, I. (2011). Taming the elephants: New TCP slow start. *Computer Networks, 44*(9), 2092–2110.

Ha, S., Rhee, I., & Xu, L. (2008). CUBIC: A new TCP-friendly high-speed TCP variant. *ACM SIGOPS Operating System Review, 42*(5), 64–74.

Huang, J., Qian, F., Guo, Y., Zhou, Y., Xu, Q., Mao, Z. M., et al. (2013). An in-depth study of LTE: Effect of network protocol and application behavior on performance. *SIGCOMM Computer Communication Review, 43*(4), 363–374. https://doi.org/10.1145/2534169.2486006.

Ibrahim, M., Liu, H., Jawahar, M., Nguyen, V., Gruteser, M., Howard, R., Yu, B., & Bai, F. (2018). Verification: Accuracy evaluation of WiFi fine time measurements on an open platform. In *Proceedings of the 24th Annual International Conference on Mobile Computing and Networking (Mobicom'18)* (pp. 417–427). New York, NY, USA: Association for Computing Machinery. https://doi.org/10.1145/3241539.3241555.

Information Sciences Institute. (1981). Transmission control protocol. IETF RFC 793. https://datatracker.ietf.org/doc/html/rfc793.

Jiang, J., Sekar, V., & Zhang, H. (2012). Improving fairness, efficiency, and stability in HTTP-based adaptive video streaming with FESTIVE. In *Proceedings of the 8th International Conference on Emerging Networking Experiments and Technologies (CoNEXT '12)* (pp. 97–108). New York, NY, USA: Association for Computing Machinery.

Judd, G., Wang, X., & Steenkiste, P. (2007). Low-overhead channel-aware rate adaptation. In *Proceedings of the 13th Annual ACM International Conference on Mobile Computing and Networking (MobiCom '07)* (pp. 354–357). New York, NY, USA: Association for Computing Machinery. https://doi.org/10.1145/1287853.1287903.

Jun, B., Bustamante, F. E., Whang, S. Y., & Bischof, Z. S. (2019). AMP up your mobile web experience: Characterizing the impact of Google's accelerated mobile project. In *The 25th Annual International Conference on Mobile Computing and Networking (MobiCom '19)* (Article 4, p. 14). New York, NY, USA: Association for Computing Machinery.

Khan, I., Ghoshal, M., Aggarwal, S., Koutsonikolas, D., & Widmer, J. (2022). Multipath TCP in smartphones equipped with millimeter wave radios. In *Proceedings of the 15th ACM Workshop on Wireless Network Testbeds, Experimental Evaluation & CHaracterization (WiNTECH'21)* (pp. 54–60). New York, NY, USA: Association for Computing Machinery. https://doi.org/10.1145/3477086.3480839.

Kistler, J. J., & Satyanarayanan, M. (1992). Disconnected operation in the coda file system. *ACM Transactions on Computer Systems, 10*(1), 3–25. https://doi.org/10.1145/146941.146942.

Koopman, P. (2002). 32-Bit cyclic redundancy codes for internet Applications. *The International Conference on Dependable Systems and Networks (DSN)* (pp. 1–10). Washington, DC, USA: IEEE.

Kung, H. T., Blackwell, T., & Chapman, A. (1994). Credit-based flow control for ATM networks: Credit update protocol, adaptive credit allocation and statistical multiplexing. In *Proceedings of the Conference on Communications Architectures, Protocols and Applications (SIGCOMM '94)* (pp. 101–114). New York, NY, USA: Association for Computing Machinery. https://doi.org/10.1145/190314.190324.

Kurose, J., & Ross, K. (2021). *Computer networking* (8th ed.). London, UK: Pearson.

Langley, A., Riddoch, A., Wilk, A., Vicente, A., Krasic, C., Zhang, D., Yang, F., Kouranov, F., Swett, I., Iyengar, J., Bailey, J., Dorfman, J., Roskind, J., Kulik, J., Westin, P., Tenneti, R., Shade, R., Hamilton, R., Vasiliev, V., Chang, W.-T., & Shi, Z. (2017). The QUIC transport protocol: Design and internet-scale deployment. In *Proceedings of the Conference of the ACM Special Interest Group on Data Communication (SIGCOMM '17)* (pp. 183–196). New York, NY, USA: Association for Computing Machinery. https://doi.org/10.1145/3098822.3098842.

Le Boudec, J.-Y. (2005). Rate adaptation, congestion control and fairness: A tutorial. http://ica1www.epfl.ch/PS_files/LEB3132.pdf#search=%22max-min%20fairness%22.

Lee, H. J., Flinn, J., & Tonshal, B. (2018). RAVEN: Improving interactive latency for the connected car. In *Proceedings of the 24th Annual International Conference on Mobile Computing and Networking (MobiCom '18)* (pp. 557–572). New York, NY, USA: Association for Computing Machinery.

Li, C., & Cao, Z. (2022). LoRa networking techniques for large-scale and long-term IoT: A down-to-top survey. *ACM Computing Surveys,55*(3), Article 52, 36. https://doi.org/10.1145/3494673.

Li, L., Xu, K., Li, T., Zheng, K., Peng, C., Wang, D., Wang, X., Shen, M., & Mijumbi, R. (2018). A measurement study on multi-path TCP with multiple cellular carriers on high speed rails. In *Proceedings of the 2018 Conference of the ACM Special Interest Group on Data Communication*

(SIGCOMM '18) (pp. 161–175). New York, NY, USA: Association for Computing Machinery. https://doi.org/10.1145/3230543.3230556.

Lim, Y.-S., Chen, Y.-C., Nahum, E., Towsley, D., & Lee, K.-W. (2014). Cross-layer path management in multi-path transport protocol for mobile devices. In *Proceedings - IEEE INFOCOM* (pp. 1815–1823). Washington, DC, USA: IEEE Computer Society Press. https://doi.org/10.1109/INFOCOM.2014.6848120.

Li, L., Xu, K., Wang, D., Peng, C., Zheng, K., Mijumbi, R., & Xiao, Q. (2017). A longitudinal measurement study of TCP performance and behavior in 3G/4G networks over high speed rails. *IEEE/ACM Transactions on Networking, 25*(4), 2195–2208. https://doi.org/10.1109/TNET.2017.2689824.

Mathis, M., Semke, J., Mahdavi, J., & Ott, T. (1997). The macroscopic behavior of the TCP congestion avoidance algorithm. *SIGCOMM Computer Communication Review, 27*(3), 67–82. https://doi.org/10.1145/263932.264023.

Mishra, A., Sun, X., Jain, A., Pande, S., Joshi, R., & Leong, B. (2020). The great internet TCP congestion control census. In *Abstracts of the 2020 SIGMETRICS/Performance Joint International Conference on Measurement and Modeling of Computer Systems (SIGMETRICS '20)* (pp. 59–60). New York, NY, USA: Association for Computing Machinery. https://doi.org/10.1145/3393691.3394221.

Molisch, A. (2011). *Wireless communications*. Hoboken, New Jersey: Wiley.

Moon, T. (2005). *Error correction coding: Mathematical methods and algorithms*. Hoboken, New Jersey: Wiley Online Library.

Mukerjee, M. K., Bozkurt, I. N., Ray, D., Maggs, B. M., Seshan, S., & Zhang, H. (2017). Redesigning CDN-broker interactions for improved content delivery. In *Proceedings of the 13th International Conference on Emerging Networking EXperiments and Technologies (CoNEXT '17)* (pp. 68–80). New York, NY, USA: Association for Computing Machinery.

Murray, D. G., Schwarzkopf, M., Smowton, C., Smith, S., Madhavapeddy, A., & Hand, S. (2011). CIEL: A universal execution engine for distributed data-flow computing. In *Proceedings of the 8th USENIX Conference on Networked Systems Design and Implementation (NSDI'11)* (pp. 113–126). USA: USENIX Association.

Nagle, J. (1984). Congestion control in IP/TCP internetworks. IETF RFC 896. https://datatracker.ietf.org/doc/html/rfc896.

National Telecommunications and Information Administration. (2016). United States Frequency Allocatoin Chart. https://www.ntia.doc.gov/page/2011/united-states-frequency-allocation-chart.

Nitsche, T., Cordeiro, C., Flores, A. B., Knightly, E. W., Perahia, E., & Widmer, J. C. (2014). IEEE 802.11ad: Directional 60 GHz communication for multi-Gigabit-per-second Wi-Fi [Invited Paper]. *IEEE Communications Magazine,52*(12), 132–141.

Noble, B. D., Satyanarayanan, M., Narayanan, D., Tilton, J. E., Flinn, J., & Walker, K. R. (1997). Agile application-aware adaptation for mobility. In *Proceedings of the Sixteenth ACM Symposium on Operating Systems Principles (SOSP '97)* (pp. 276–287). New York, NY, USA: Association for Computing Machinery. https://doi.org/10.1145/268998.266708.

Nygren, E., Sitaraman, R. K., & Sun, J. (2010). The Akamai network: A platform for high-performance internet applications. *SIGOPS Operating Systems Review, 44*(3), 2–19.

Paasch, C., Detal, G., Duchene, F., Raiciu, C., & Bonaventure, O. (2012). Exploring Mobile/WiFi handover with multipath TCP. In *Proceedings of the 2012 ACM SIGCOMM Workshop on Cellular Networks: Operations, Challenges, and Future Design (CellNet '12)* (pp. 31–36). New York, NY, USA: Association for Computing Machinery. https://doi.org/10.1145/2342468.2342476.

Padhye, J., Firoiu, V., Towsley, D., & Kurose, J. (1998). Modeling TCP throughput: A simple model and its empirical validation. In *Proceedings of the ACM SIGCOMM '98 Conference on Applications, Technologies, Architectures, and Protocols for Computer Communication (SIGCOMM '98)* (pp.

303–314). New York, NY, USA: Association for Computing Machinery. https://doi.org/10.1145/285237.285291.

Pan, Y., Li, R., & Xu, C. (2022). The first 5G-LTE comparative study in extreme mobility. *Proceedings of the ACM on Measurement and Analysis of Computing Systems,6*(1), Article 20, 22. https://doi.org/10.1145/3508040.

Pei, C., Wang, Z., Zhao, Y., Wang, Z., Meng, Y., Pei, D., Peng, Y., Tang, W., & Qu, X. (2017a). Why it takes so long to connect to a WiFi access point. In *IEEE INFOCOM 2017 - IEEE Conference on Computer Communications* (pp. 1–9). IEEE. https://doi.org/10.1109/INFOCOM.2017.8057164.

Pei, C., Wang, Z., Zhao, Y., Wang, Z., Meng, Y., Pei, D., Peng, Y., Tang, W., & Qu, X. (2017b). Why it takes so long to connect to a WiFi access point. In *IEEE INFOCOM 2017 - IEEE Conference on Computer Communications* (pp. 1–9). Piscataway, NJ: IEEE. https://doi.org/10.1109/INFOCOM.2017.8057164.

Perkins, C. (1996). IP mobility support. IETF RFC 2002. https://datatracker.ietf.org/doc/html/rfc2002.

Perkins, C. (1997). Mobile IP. *IEEE Communications Magazine, 35*(5), 84–99.

Peterson, L., & Davie, B. (2020). Computer networks: A systems approach. https://book.systemsapproach.org/.

Philip, A. A., Ware, R., Athapathu, R., Sherry, J., & Sekar, V. (2021). Revisiting TCP congestion control throughput models & fairness properties at scale. In *Proceedings of the 21st ACM Internet Measurement Conference* (pp. 96–103). New York, NY, USA: Association for Computing Machinery. https://doi.org/10.1145/3487552.3487834.

Ra, M.-R., Sheth, A., Mummert, L., Pillai, P., Wetherall, D., & Govindan, R. (2011). Odessa: Enabling interactive perception applications on mobile devices. In *Proceedings of the 9th International Conference on Mobile Systems, Applications, and Services (MobiSys '11)* (pp. 43–56). New York, NY, USA: Association for Computing Machinery.

Ramakrishnan, K., Floyd, S., & Black, D. (2001). The addition of Explicit Congestion Notification (ECN) to IP. IETF RFC 3168. https://datatracker.ietf.org/doc/html/rfc3168.

Rescorla, E. (2018). The Transport Layer Security (TLS) protocol version 1.3. IETF RFC 8446. https://datatracker.ietf.org/doc/html/rfc8446.

Rhee, I., Xu, L., Ha, S., Zimmermann, A., Eggert, L., & Scheffenegger, R. (2018). CUBIC for fast long-distance networks. IETF RFC 8312. https://tools.ietf.org/html/rfc8312.

Rhee, I., Xu, L., Ha, S., Zimmermann, A., Eggert, L., & Scheffenegger, R. (2022). Ceilo Networks. https://www.cielonetworks.com.

Riggio, R., Marina, M. K., Schulz-Zander, J., Kuklinski, S., & Rasheed, T. (2015). Programming abstractions for software-defined wireless networks. *IEEE Transactions on Network and Service Management, 12*(2), 146–162.

Roberts, L. (1986). The arpanet and computer networks. In *Proceedings of the ACM Conference on The History of Personal Workstations (HPW '86)* (pp. 51–58). New York, NY, USA: Association for Computing Machinery. https://doi.org/10.1145/12178.12182.

Rubens, A., Rigney, C., Willens, S., & Simpson, W. (2000). Remote authentication dial in user service (RADIUS). IETF RFC 2865. https://tools.ietf.org/html/rfc2865.

Saltzer, J. H., Reed, D. P., & Clark, D. D. (1984). End-to-end arguments in system design. *ACM Transactions on Computer Systems, 2*(4), 277–288. https://doi.org/10.1145/357401.357402.

Satyanarayanan, M. (1990). Scalable, secure, and highly available distributed file access. *Computer,23*(5), 9–18, 20–21.

Satyanarayanan, M. (2017). The emergence of edge computing. *IEEE Computer, 50*(1), 30–39.

Satyanarayanan, M., Bahl, B., Caceres, R., & Davie, S. N. (2009). The case for VM-based cloudlets in mobile computing. *IEEE Pervasive Computing, 8*(4), 14–23.

Schomp, K., Bhardwaj, O., Kurdoglu, E., Muhaimen, M., & Sitaraman, R. K. (2020). Akamai DNS: Providing authoritative answers to the world's queries. In *Proceedings of the Annual Conference of the ACM Special Interest Group on Data Communication on the Applications, Technologies, Architectures, and Protocols for Computer Communication (SIGCOMM '20)* (pp. 465–478). New York, NY, USA: Association for Computing Machinery.

Schwartz, M., & Abramson, N. (2009). The alohanet - surfing for wireless data [History of Communications]. *Communications Magazine, 47*(12), 21–25. https://doi.org/10.1109/MCOM.2009.5350363.

Shrivastava, V., Rayanchu, S., Yoonj, J., & Banerjee, S. (2008). 802.11n under the microscope. In *Proceedings of the 8th ACM SIGCOMM Conference on Internet Measurement (IMC '08)* (pp. 105–110). New York, NY, USA: Association for Computing Machinery. https://doi.org/10.1145/1452520.1452533.

Sivakumar, A., Jiang, C., Nam, Y. S., Puzhavakath Narayanan, S., Gopalakrishnan, V., Rao, S. G., Sen, S., Thottethodi, M., & Vijaykumar, T. N. (2017). NutShell: Scalable whittled proxy execution for low-latency Web over cellular networks. In *Proceedings of the 23rd Annual International Conference on Mobile Computing and Networking (MobiCom '17)* (pp. 448–461). New York, NY, USA: Association for Computing Machinery.

Snoeren, A., & Balakrishna, H. (2000). An end-to-end approach to host mobility. In *Proceedings of the 6th Annual International Conference on Mobile Computing and Networking (MobiCom'00)* (pp. 155–166). New York, NY, USA: Association for Computing Machinery.

Spencer, Q. H., Peel, C. B., Swindlehurst, A. L., & Haardt, M. (2004). An introduction to the multi-user MIMO downlink. *Communications Magazine, 42*(10), 60–67. https://doi.org/10.1109/MCOM.2004.1341262.

Stenberg, D. (2014b). HTTP2 explained. *SIGCOMM Computer Communication Review, 44*(3), 120–128.

Stenberg, D. (2014a). HTTP2 explained. *SIGCOMM Computer Communication Review, 44*(3), 120–128.

Stevens, W. (1997). TCP slow start, congestion avoidance, fast retransmit, and fast recovery. IETF RFC 2001. https://datatracker.ietf.org/doc/html/rfc2001.

Sui, K., Zhou, M., Liu, D., Ma, M., Pei, D., Zhao, Y., Li, Z., & Moscibroda, T. (2016). Characterizing and improving WiFi latency in large-scale operational networks. In *Proceedings of the 14th Annual International Conference on Mobile Systems, Applications, and Services (MobiSys '16)* (pp. 347–360). New York, NY, USA: Association for Computing Machinery. https://doi.org/10.1145/2906388.2906393.

Sun, W., Choi, M., & Choi, S. (2013). IEEE 802.11ah: A long range 802.11 WLAN at Sub 1 GHz. *Journal for ICT Standardization,1*(1), 83–108.

Tan, Z., Li, Z., Li, Y., Xu, Y., & Lu, S. (2021). Device-based LTE latency reduction at the application layer. In *18th USENIX Symposium on Networked Systems Design and Implementation (NSDI 21)* (pp. 471–486). Usenix Association, virtual.

Tananbaum, A., Feamster, N., & Wetherall, D. (2021). *Computer networking* (6th ed.). London, UK: Pearson.

Thakur, N. R., & Kunte, A. S. (2020). Analysis of MPTCP packet scheduling, the need of data hungry applications. In G. Ranganathan, J. Chen, & Á. Rocha (Eds.), *Inventive Communication and Computational Technologies* (pp. 599–617). Singapore: Springer Singapore.

Vollbrecht, J., Carlson, J., Blunk, L., Aboba, B., & Levkowetz, H. (2004). Extensible Authentication Protocol (EAP). IETF RFC 3748. https://tools.ietf.org/html/rfc3748.

Vu, V. A., & Walker, B. (2019). Redundant multipath-TCP scheduling with desired packet latency. In *Proceedings of the 14th Workshop on Challenged Networks (CHANTS'19)* (pp. 7–12). New York, NY, USA: Association for Computing Machinery. https://doi.org/10.1145/3349625.3355440.

Wang, S., Huang, J., & Zhang, X. (2020). Demystifying millimeter-wave V2X: Towards robust and efficient directional Connectivity under high mobility. In *Proceedings of the 26th Annual International Conference on Mobile Computing and Networking* (Article 51, p. 14). New York, NY, USA: Association for Computing Machinery.

Wang, J., Zheng, Y., Ni, Y., Xu, C., Qian, F., Li, W., Jiang, W., Cheng, Y., Cheng, Z., Li, Y., Xie, X., Sun, Y., & Wang, Z. (2019). An active-passive measurement study of TCP performance over LTE on high-speed rails. In *The 25th Annual International Conference on Mobile Computing and Networking (MobiCom '19)* (Article 18, p. 16). New York, NY, USA: Association for Computing Machinery. https://doi.org/10.1145/3300061.3300123.

Wischik, D., Raiciu, C., Greenhalgh, A., & Handley, M. (2011). Design, implementation and evaluation of congestion control for multipath TCP. In *Proceedings of the 8th USENIX Conference on Networked Systems Design and Implementation (NSDI'11)* (pp. 99–112). USA: USENIX Association.

Wright, K.-L., Sivakumar, A., Steenkiste, P., Yu, B., & Bai, F. (2020). CloudSLAM: Edge offloading of stateful vehicular applications. In *2020 IEEE/ACM Symposium on Edge Computing (SEC)* (pp.139–151). Piscataway, NJ: IEEE. https://doi.org/10.1109/SEC50012.2020.00018.

Zhang, Z., Shi, S., Gupta, V., & Jana, R. (2019). Analysis of cellular network latency for edge-based remote rendering streaming applications. In *Proceedings of the ACM SIGCOMM 2019 Workshop on Networking for Emerging Applications and Technologies (NEAT'19)* (pp. 8–14). New York, NY, USA: Association for Computing Machinery. https://doi.org/10.1145/3341558.3342199.